Learning
MATLAB

Learning
MATLAB

Tobin A. Driscoll
University of Delaware
Newark, Delaware

siam ® Society for Industrial and Applied Mathematics

Library of Congress Cataloging-in-Publication Data

Driscoll, Tobin A. (Tobin Allen), 1969-
  Learning MATLAB / Tobin A. Driscoll.
      p. cm.
  Includes index.
  ISBN 978-0-898716-83-2
  1. Mathematics--Data processing. 2. MATLAB. I. Title.
  QA76.95.D75 2009
  620.001'51--dc22   2009008516

To Jen, Luke, and Adam,
who are my greatest teachers.

Contents

List of Figures

List of Tables

Preface

The purpose of this book is to introduce the essentials of the MATLAB® software environment and to show how to start using it well.

MATLAB began life as a friendly interface to numerical libraries for linear algebra. Every variable in MATLAB was a matrix, which made it easy to learn how to solve certain core problems and interact with the results. Over time, as interest in MATLAB shifted from pedagogy to larger and more complex applications, the limitations and annoyances of programming with only a text-based interface to matrices became apparent. In accordance, MATLAB added higher dimensionality, many more data types, graphical and object-oriented interfaces, and loads of additional technical and helper functions.

The result has been spectacularly, perhaps uniquely, successful both commercially and in terms of educational influence. Few successes are unqualified, however, and one price of the success of MATLAB has been the loss of its initial simplicity and unity of concept. Upon starting MATLAB version 4 in 1992, one got a simple prompt demanding that the user start defining variables and running functions on them. Upon starting MATLAB 7 in 2008, one gets four tabbed windows, six main menus, and a dozen or so clickable buttons. The prompt is still there, but it is now just one familiar face in a crowd.

This observation is not meant as a Luddite screed against the perils of progress! Yet complexity *has* made the job of getting to know MATLAB more difficult—or so it would seem, given the cornucopia of books now available that have as their primary purpose the aim of teaching it.

Why do I add another book to this pile? Mainly because of the thickness, or rather the slenderness, of the volume you now hold. I offer you a guide on a human scale, rather than the scale of MATLAB. One of the jobs of this book is to get out of the way in a timely manner. You won't find a lot of application areas in this guide; nor will you find mathematical properties or descriptions of numerical algorithms. You will find, in places, some practical but editorial advice based on my 16 years of MATLAB programming experience.

You wouldn't expect to learn how to play piano without sitting at a keyboard. The same is true for learning MATLAB! You are well advised to enter the examples yourself, read online help about a newly introduced function, and, above all, try the exercises. Because I do assume that you are an active reader, I list some commands simply by name and with a brief description, with the idea that you will look up the details online if interested.

The first four chapters cover the most essential material. Chapter 5 is a bare-bones introduction to the big subject of graphics, which is also fairly indispensable. Chapter 6 covers some advanced topics, and Chapter 7 introduces commands useful in scientific computation. The version of MATLAB referred to in this book is 7.7 (Release 2008b). Virtually everything presented will work in older versions, but the probability of finding exceptions increases as the version number decreases, particularly if it gets below 7.

No prior knowledge of MATLAB should be necessary to read this book, though a working knowledge of programming is a big help. Mathematically, you should have a fair understanding of calculus and the mechanics of matrix algebra. If some of the linear algebra terminology is unfamiliar to you, it can probably be safely skipped until you have seen the relevant mathematics elsewhere.

This book could not have come to be without a lot of help. In particular I must thank Nick Trefethen, who made many instructive comments, and an anonymous reviewer, who did likewise. Thanks go especially to the Mathematical Sciences Department at the University of Delaware, for giving graduate students the opportunity to take a summer workshop on MATLAB, and me the opportunity to teach it since 2001. (The workshop was initially partially supported by a Group Infrastructure Grant from the National Science Foundation.) The Delaware students, who persevered through early drafts and suffered as test subjects for difficult exercises, deserve credit as well. Finally, I give thanks to my wife Jen, who never once showed anything less than enthusiasm and support for a book about MATLAB.

Toby Driscoll

Chapter 1

Introduction

The MATLAB software package is used for computation in engineering, science, and applied mathematics. It offers a powerful programming language, excellent graphics, and a large standard library.

The focus in MATLAB is on computation, not mathematics: symbolic expressions and manipulations are not possible, except through the optional Symbolic Toolbox, which is not covered in this book. All variables must have values, and all results are numerical and potentially inexact, thanks to the rounding errors inherent in computer arithmetic. The emphasis on numerics is typical for most work in scientific computation.

Compared to other numerically oriented languages, such as C++ and Fortran, MATLAB is usually found to be much easier to use. However, its execution speed can be slower. This gap is not always as dramatic as popular lore has it, and it can often be narrowed or closed with good MATLAB programming. Moreover, one can link other types of code into MATLAB, or vice versa, and MATLAB has some optional support for parallel computing. Still, MATLAB is usually not the tool of choice for high-performance computing.

Whatever you think of these or other limitations of MATLAB, they have not held back its popularity: a recent search for "matlab" on the books section of Amazon.com turned up 8,543 results! Rapid code development and interaction with data often trump execution speed, and the integrated graphics and expert routines that come with MATLAB can be decisively helpful. Even for speed-hungry users, MATLAB can be a valuable environment in which to explore and fine-tune algorithms before creating production code in another environment.

Successful computing languages and environments reflect a distinctive set of values. In MATLAB, those values include an emphasis on experimentation and interaction with data and algorithms; syntax that is compact, friendly, and interactive (rather than tightly constrained and verbose); a kitchen-sink mentality for providing functionality; and a predilection for vectors, matrices, and arrays.

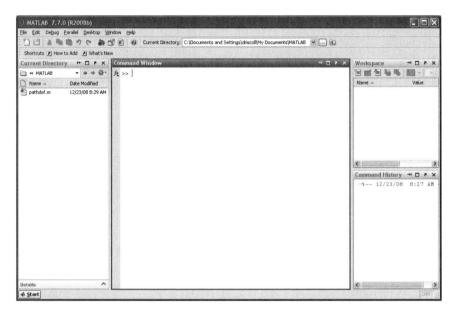

Figure 1.1. *Initial MATLAB desktop window. (Microsoft® Windows® XP version shown.)*

1.1   The fifty-cent tour

When you start MATLAB, you get a multipaneled desktop, as seen in Figure 1.1. The layout and behavior of the desktop and its components are highly customizable. The component that is the heart of MATLAB is called the **Command Window**, located in the middle by default. Here you can give MATLAB commands typed at the **prompt**, shown as >>. Unlike Fortran and other compiled computer languages, MATLAB is an interpreted environment—you give a command, and MATLAB tries to execute it right away, then awaits another.

At the left you can see the **Current Directory** window. In general MATLAB is aware only of files in the current directory (folder) and on the list known as its **path**, which can be customized. Commands for working with the directory and path include cd, what, which, addpath, and editpath (or you can choose Set path ... from the File menu). You can add files to a directory on the path and thereby add commands to MATLAB; we will return to this subject in Chapter 3.

At the top right is the **Workspace** window. The Workspace shows you what variable names are currently defined and some information about their contents. At start-up it is, naturally, empty. This represents another break from compiled environments: variables created in the workspace persist for you to examine and modify, even after code execution stops.

Below the Workspace window is the **Command History** window. As you enter commands, they are recorded here. This record persists across different MATLAB sessions, and commands or blocks of commands can be copied from here or saved to files. Thus the Command History is very useful if you realize belatedly that you need to save some or all of what you have been doing interactively.

As you explore MATLAB, you may encounter some **toolboxes**. These are individually packaged sets of capabilities that provide in-depth expertise on particular subject areas. There is no need to issue a command to load them—once installed, they appear on the path and are available automatically.

1.2 Graphical versus command-line usage

Originally, MATLAB was entirely a command-line environment, and it retains a bias in that direction.[1] But it is possible to access a great deal of the functionality from graphical widgets such as menus, buttons, and so on. These interfaces are especially useful to beginners, because they lay out the available choices clearly. In particular, take time to right-click (or Control-click on a Mac®) on various objects to see what you might be able to do with them.

As a rule, graphical interfaces can be more natural for certain types of interactive work, such as annotating a single graph, whereas typed commands remain better for complex, precise, repeated, or reproducible tasks. Much of the time, you can choose whichever mode of operation suits you. For instance, you can write a function that customizes any figure's appearance, but you can also save aspects of a current figure's style as a template just by pointing and clicking. Moreover, you can create your own graphical interfaces and even distribute them with your code as a package for non-MATLAB users. In the end, an advanced MATLAB user should be able to exploit both modes of operation. That said, the focus of this book is on typed commands. In many, perhaps most, cases these have graphical interface equivalents, even if I don't explicitly point them out.

1.3 Help

MATLAB is huge. Nobody can tell you everything that you personally will need to know; nor could you remember it all anyway. It is essential that you become familiar with the online help. There are two levels of help:

- If you need quick help on the syntax of a command, type `help` in the command window. For example, `help plot` shows directly in the Command Window all the ways in which you can use the `plot` command. Typing `help` by itself gives you a list of categories that themselves yield lists of commands.

- Typing `doc` followed by a command name brings up more extensive help in a separate window. For example, `doc plot` is better formatted and more informative than `help plot`. In the left panel one sees a hierarchical, browsable display of all the online documentation. Typing `doc` alone or

---

[1]You can still run MATLAB entirely in this mode by starting it with the `-nojvm` option, which can be convenient when connecting to a remote server, for example.

selecting Help from the drop-down menus brings up the window at a root homepage.

Starting in MATLAB 7.7, the Command Window includes a **function browser** whose icon is *fx*, sitting next to the prompt. Clicking on it yields a searchable, hierarchical listing of available functions, with help available on one more click. Merely pausing after typing the name of a function and its opening parenthesis brings up a context-sensitive box with syntactic suggestions, as well.

The *Getting Started with MATLAB* manual is a good place to get a more leisurely and thorough introduction than the one to follow here. Depending on your installation, the documentation may be available in PDF form for printing and offline reading, or you can find it on the Web at: www.mathworks.com/access/ helpdesk/help/pdf_doc/matlab/getstart.pdf

Complementary to the online help is MATLAB Central, found on the Web at www.mathworks.com/matlabcentral. It includes a discussion forum and the File Exchange, which consists of code contributions by MATLAB users and friends. It's a good place to check when you suspect that you may be trying to reinvent the wheel.

1.4   Basic commands and syntax

If you type in a valid expression and press Enter, MATLAB will immediately execute it and return the result, just like a calculator.

```
>> 2+2
ans =
      4

>> (6.02e23)^2
ans =
  3.6240e+047

>> sin(pi/2)
ans =
      1

>> 1/0

ans =
   Inf

>> exp(i*pi)
ans =
  -1.0000 + 0.0000i
```

Notice some of the special expressions here: pi for π, Inf for ∞, and i for $\sqrt{-1}$. You may also use j for the imaginary unit. (Both names can be reassigned, however, and are commonly used as integers. It's safer always to use 1i or 1j to refer to the imaginary unit.) Another special value is NaN, which stands for **not a number**. NaN is used to express an undefined value. For example,

```
>> Inf/Inf
ans =
    NaN
```

NaN can be tricky to use: two NaN values are unequal by definition, for instance. You can assign values to variables with alphanumeric names:

```
>> x = sqrt(3)
x =
    1.7321

>> days_since_birth = floor(now) - datenum(1969,05,06)
days_since_birth =
        12810

>> 3*z
??? Undefined function or variable 'z'.
```

Observe that *variables must have values before they can be used.* When an expression returns a single result that is not assigned to a variable, this result is assigned to ans, which can then be used like any other variable:

```
>> atan(x)
ans =
    1.0472

>> pi/ans
ans =
    3
```

In floating-point arithmetic, you should not expect "equal" values to have a difference of exactly zero. The built-in number eps, often called **unit roundoff** or **machine precision** in numerical analysis, bounds the maximum relative error in representing real numbers and doing arithmetic on your particular machine.[2]

---

[2]Like all other names, eps can be reassigned, but doing so has no effect on the precision.

```
>> eps
ans =
   2.2204e-16

>> exp(log(10)) - 10
ans =
   1.7764e-15

>> ans/10
ans =
   1.7764e-16
```

Here are a few other demonstration statements.

```
>> % Anything after a % sign is a comment.
>> x = rand(100,100); % ; to suppress output
>> s = 'Hello world'; % single quotes enclose a string
>> t = 1 + 2 + 3 + ...
4 + 5 + 6                  % ... continues a line

t =
    21
```

Once variables have been defined, they exist in the workspace. You can see what's in the workspace from its desktop window, or by typing who or whos:

```
>> who

Your variables are:

ans   s    t    x
```

1.5   Saving and loading work

If you enter save myfile, all defined variables will be saved to a file called myfile.mat in the current directory. This file format is particular to MATLAB. You can also select a subset of variables to be saved by typing their names after the file name. If you later enter load myfile, the saved variables are returned to the workspace, overwriting any presently defined values assigned to the same names.

If you highlight commands in the Command History window, right-click, and select Create M-File you can save the typed commands to a text file. This can be very helpful for recreating what you have done. Another way to save both input and output is to use the diary command with a file name. This causes all subsequent commands and results to be recorded in a text file on disk. A more sophisticated version of the same idea is called **publishing**, which is discussed in section 3.1 on page 28.

MATLAB is also capable of loading and saving other common file formats, such as formatted text files, spreadsheet files, and common graphics and video files. To load such a file, it's often easiest to double-click it in the Current Directory window, and follow the resulting prompts.

1.6 Things about MATLAB that are very nice to know, but which often do not come to the attention of beginners

Don't worry if everything in this list doesn't make sense yet—come back to it again later.

(a) Use the up-arrow key to cycle through previous commands. If you type specific characters first, only commands matching the typed characters will be recalled.

(b) If a computation is taking too long, interrupt it by pressing Ctrl-C (after making sure the Command Window is active in the operating system).

(c) Even when MATLAB displays only 4–5 digits of a result, it's storing about 15 significant digits. (You can see them all by typing `format long`.) By copying or retyping a displayed result, you throw away a lot of information. Wait until the end of the calculation to round off results.

(d) MATLAB has great debugging tools. Run your code step by step to uncover errors. Run someone else's code step by step to understand it thoroughly.

(e) The M-Lint code checker makes some good suggestions.

(f) The previous two items alone are sufficient reasons to use the built-in MATLAB Editor for writing code. Open it by entering `edit`.

(g) If the execution of your code is too slow to suit, use the Profiler to find the slowest steps. Colleagues will be amazed at how you always home in on the bottlenecks.

(h) If you want to share code or use it on a different machine, first use the `depfun` command on it to find the files it depends on. Or, check out the section of online help on "directory reports."

(i) Don't use a screen or window capture function to paste figures into a document or presentation. The results look cheesy and amateurish. Instead, have a look at section 5.5.

(j) After you have properly exported a figure to a graphics file, save that figure again in the native FIG format. You may want to make changes to it someday.

(k) Plot legends are helpful and can be generated automatically. Better still are curves with labels right next to them—also very easy to create.

(l) Anonymous functions are really powerful. Trust me on this, and go learn about them, beginning in section 4.1.

Exercises

1.1. Evaluate the following mathematical expressions in MATLAB:

(a) $\tanh(e)$ (b) $\log_{10}(2)$ (c) $\log_2(10)$

(d) $\left|\sin^{-1}\left(-\frac{1}{2}\right)\right|$ (e) 123456 mod 789 (f) $\operatorname{Arg}(1+i\sqrt{2})$

1.2. MATLAB ships with some interesting data. For example, try the following:

```
load usapolygon, plot(uslon,uslat)
```

Use who or the Workspace browser to find out where the data comes from.

1.3. What is the name of the built-in function that MATLAB uses to
1. compute a Bessel function of the second kind?
2. test the primality of an integer?
3. multiply two polynomials together?
4. plot a vector field?
5. report the current date and time?

1.4. Find a function on the MATLAB Central File Exchange that enables you to delete the most recently created graphics object from the command line.

1.5. MATLAB ships with some built-in capability for working with images (especially if you have the Image Toolbox available). After all, an image is usually represented as either one array or three arrays of pixel intensity values. Use the commands

```
>> load durer
>> image(X)
>> colormap(map)
```

to see a reproduction of a mathematically themed work of art. Then use the online documentation to find out more about the matrix that appears in this artwork.

About the MATLAB Treasure Hunt on the next page: I load a briefcase with a reward and secure it using a three-digit luggage lock. Teams or individuals work to find the last answer in the hunt and try it as the combination of the lock. The first team to open the lock keeps the bounty.

A MATLAB Treasure Hunt

Follow the directions. You may use only MATLAB and its local online help.

Find the largest prime factor of 20830123: $\alpha = $ _____

Find the complete elliptic integral of the first kind, $K(1 - 1/\alpha^2)$, rounded to three significant digits: $\beta = $ _____

Find the remainder after the largest possible positive 32-bit integer in MATLAB is divided by 100β: $\gamma = $ _____

Find the maximum element in a $\gamma \times \gamma$ symmetric Clement matrix, rounded to the nearest integer: $\delta = $ _____

Find the number of minutes that elapsed between January 20, 1961 at 12:51 PM, and July 16, 1969 at 9:32 AM. Divide by 100δ, and round to the nearest integer: $\epsilon = $ _____

Chapter 2

Arrays and Matrices

The heart and soul of the MATLAB software is linear algebra. In fact, "MAT-LAB" was originally a contraction of "matrix laboratory." More so than any other language, MATLAB encourages and expects you to make heavy use of arrays, vectors, and matrices.

Some jargon: An **array** is a collection of numbers, called **elements** or **entries**, referenced by one or more indices running over different index sets. In MATLAB, the index sets are always sequential integers starting with 1. The **dimension** of the array is the number of indices needed to specify an element. The **size** of an array is a list of the sizes of the index sets.

A **matrix** is a two-dimensional array with special rules for addition, multiplication, and other operations. It represents a mathematical linear transformation. The two dimensions are called **rows** and **columns**. A **vector** is a matrix for which one dimension has only the index 1: a **row vector** has only one row, and a **column vector** has only one column.

Although an array is more general and less mathematical than a matrix, the terms are often used interchangeably. What's more, in MATLAB there is really no formal distinction. The commands in this chapter are sorted according to the array/matrix distinction, but MATLAB will let you mix them freely as long as the syntax is defined. The idea—here, as elsewhere—is that MATLAB keeps the language simple, natural, and succinct. It's up to you to stay out of trouble.

2.1   Building arrays and matrices

The most straightforward way to construct a small array is by enclosing its elements in square brackets. Use spaces or commas to separate columns, and use semicolons or new lines to separate rows:

```
>> A = [1 2 3; 4 5 6; 7 8 9]
A =
         1        2        3
         4        5        6
         7        8        9

>> b = [0;1;0]
b =
         0
         1
         0
```

Information about size and dimension is stored with the array:[3]

```
>> size(A)
ans =
         3        3

>> size(b)
ans =
         3        1

>> nd = [ ndims(A), ndims(b) ]
nd =
         2        2
```

Notice that there is really no such thing as a one-dimensional array in MATLAB. Vectors are technically two-dimensional, with a trivial dimension. The distinction between row and column vectors is often, but not always, important. Table 2.1 lists more commands for obtaining information about an array.

Table 2.1. *Matrix/array information commands.*

size	size in each dimension
length	size of longest dimension (especially for vectors)
ndims	number of dimensions
find	indices of nonzero elements

---

[3] Because of this, array sizes are not usually passed explicitly to functions as they are in Fortran.

Arrays can be built out of other arrays, as long as the sizes are compatible:

```
>> [A b]
ans =
     1     2     3     0
     4     5     6     1
     7     8     9     0

>> [A;b]
??? Error using ==> vertcat
All rows in the bracketed expression must have the
same number of columns.

>> B = [ [1 2;3 4] [5;6] ]
B =
     1     2     5
     3     4     6
```

One important special array is the **empty matrix**, which is entered as [].
Direct bracket constructions are suitable only for small matrices. For larger ones, there are many useful functions, some of which are shown in Table 2.2. For example:

```
>> [eye(3)   diag([1 2 3])]
ans =
     1     0     0     1     0     0
     0     1     0     0     2     0
     0     0     1     0     0     3

>> [zeros(2,3) ones(2,4)]
ans =
     0     0     0     1     1     1     1
     0     0     0     1     1     1     1
```

These and similar functions are sufficient to create most useful arrays, removing the need for a doubly nested loop in row and column indices. An especially important array constructor is the **colon** operator:

```
>> i1 = 1:8
i1 =
     1     2     3     4     5     6     7     8

>> i2 = 0:2:10
i2 =
     0     2     4     6     8    10
```

```
>> i3 = 1:-.5:-1
i3 =
    1.0000    0.5000         0   -0.5000   -1.0000
```

The format is first:step:last. The result is always a row vector, or the empty matrix if last is less than first.

Table 2.2. *Commands for building arrays and matrices.*

:	linearly spaced vector
eye	identity matrix
zeros	all zeros
ones	all ones
diag	diagonal matrix (or, extract a diagonal)
toeplitz	constant on each diagonal
triu	upper triangle
tril	lower triangle
rand, randn	random entries
linspace	linearly spaced vector
cat	concatenate along a given dimension
repmat	duplicate a vector across a dimension

2.2   Referencing elements

It is frequently necessary to access one or more of the elements of a matrix. Each dimension is given a single index or vector of indices. The result is a block extracted from the matrix. The colon is often a useful way to construct array indices. Here are some examples, using the definitions above:

```
>> A(2,3)        % single element
ans =
     6

>> b(2)          % one index for a vector
ans =
     1

>> b([1 3])      % multiple elements
ans =
     0
     0

>> A(1:2,2:3)    % a submatrix
ans =
     2     3
     5     6
```

In many situations, one wants to specify indices relative to the beginning or end of the array, or to include all possible indices. A few examples show how this is done:

```
>> B(1,2:end)           % first row, all but first column
ans =
      2      5

>> A(1:end-1,end)       % all but last row, last column
ans =
      3
      6

>> B(:,3)               % all rows of column 3
ans =
      5
      6

>> b(:,[1 1 1 1])       % multiple copies of a column
ans =
      0      0      0      0
      1      1      1      1
      0      0      0      0
```

A colon by itself as an index is understood to mean "all entries in this dimension." Observe from the last example above that an indexing result need not be a subset of the original array.

It's quite natural to access elements of a vector using a single subscript. However, any array is trivially equivalent to a vector, because it is stored linearly in memory, varying over the first dimension, then the second, and so on. (Think of the columns of an array being stacked on top of each other.) Hence a single subscript can be used for any array, automatically "flattening" it. See the documentation on sub2ind and ind2sub for more details.

```
>> A
A =
      1      2      3
      4      5      6
      7      8      9

>> A([1 2 3 4])
ans =
      1      4      7      2
```

```
>> A(:)
ans =
     1
     4
     7
     2
     5
     8
     3
     6
     9
```

The output of access by a single index is in the same shape as the index. The idiom A(:), however, is always a column vector.

Subscript referencing can be used on either side of assignments. An array is resized automatically if you delete elements or make assignments outside the current size as shown below. (Any new undefined elements are set to zero.)

```
>> C = rand(2,5)
C =
    0.2610      0.6103      0.0642      0.2259      0.0736
    0.8007      0.8741      0.1214      0.6702      0.0380

>> C(1,:) = ones(1,5)
C =
    1.0000      1.0000      1.0000      1.0000      1.0000
    0.8007      0.8741      0.1214      0.6702      0.0380

>> C(:,4) = -1      % expand scalar into the submatrix
C =
    1.0000      1.0000      1.0000     -1.0000      1.0000
    0.8007      0.8741      0.1214     -1.0000      0.0380

>> C(:,2) = []      % delete elements
C =
    1.0000      1.0000     -1.0000      1.0000
    0.8007      0.1214     -1.0000      0.0380

>> C(3,3) = 3       % grow the array and fill with zeros
C =
    1.0000      1.0000     -1.0000      1.0000
    0.8007      0.1214     -1.0000      0.0380
         0           0      3.0000           0
```

Automatic resizing and scalar expansions can be highly convenient, but they can also cause hard-to-find mistakes and even subtle performance penalties (see section 6.1).

Table 2.3. *Relational operators.*

==	equal to	˜=	not equal to
<	less than	>	greater than
<=	less than or equal to	>=	greater than or equal to

A different kind of array indexing is **logical indexing**. Logical indices usually arise from a **relational operator** (see Table 2.3). The result of applying a relational operator is a **logical array**, whose elements are 0 and 1 with interpretation as "false" and "true." Using a logical array as an index selects those values where the index is 1. A logical index can be used to select in just one dimension, or (more commonly) used as a single index in the flat-indexing model.

```
>> B = floor( 5*rand(2,4) )
B =
      0     1     2     1
      3     2     0     4

>> B>2
ans =
      0     0     0     0
      1     0     0     1

>> B(ans)
ans =
      3
      4

>> B(B==0) = NaN
B =
    NaN     1     2     1
      3     2   NaN     4
```

A less direct way of accomplishing the same thing is to use `find`, which returns the flat-index locations of nonzeros in any array, including a logical one:

```
>> find(B>2)
ans =
      2
      8

>> B(ans)
ans =
      3
      4
```

The disadvantage of `find` in this context is that you cannot easily refer to the complementary set of elements, whereas the ~ operator does complementation for logical arrays.

You can also create logical indices by hand, but you must explicitly cast them as such. Note carefully the difference between these two cases:

```
>> b = [1 2 3];
>> b([1 1 1])              % first element, three copies
ans =
       1       1       1

>> b(logical([1 1 1]))     % every element
ans =
       1       2       3
```

2.3   Matrix operations

The arithmetic operators +,-, *, ^ are interpreted in a matrix sense:

```
>> A = [1 2 3; 4 5 6; 7 8 9];
>> D = [0 0 1; 0 0 1; 0 0 1];
>> A - 3*D
ans =
       1       2       0
       4       5       3
       7       8       6

>> Asq = A^2
Asq =
      30      36      42
      66      81      96
     102     126     150
>> b = [0; 1; -1];
>> B = [1 -1 1; 1 1 1];
>> Ab = A*b
Ab =
      -1
      -1
      -1

>> AB = A*B
??? Error using ==> mtimes
Inner matrix dimensions must agree.
```

```
>> BA = B*A
BA =
        4      5      6
       12     15     18
```

The apostrophe or single-quote mark ' produces the complex-conjugate transpose (i.e., the Hermitian or adjoint) of a matrix:

```
>> zero = A*B' - (B*A')'
zero =
        0      0
        0      0
        0      0

>> inner = b'*b
inner =
        2

>> outer = b*b'
outer =
        0      0      0
        0      1     -1
        0     -1      1
```

A special operator, the backslash \, is used to solve linear systems of equations:

```
>> A = magic(3);
>> x = A\b
x =
    -0.2083
    -0.0833
     0.2917

>> b-A*x
ans =
  1.0e-015 *

          0
     0.2220
     0.2220
```

(The last result gives 10^{-15} as a factor scaling the entire vector shown.) Mathematically, when A is invertible and B is any matrix of compatible size, A\B is equivalent (given exact arithmetic) to $A^{-1}B$. Similarly, B/A is equal to BA^{-1} if defined.

Table 2.4. *Functions from linear algebra.*

\	solve linear system (or least squares)
rank	rank
det	determinant
norm	norm (2-norm, by default)
expm	matrix exponential
lu	LU factorization (Gaussian elimination)
qr	QR factorization
chol	Cholesky factorization
eig	eigenvalue decomposition
svd	singular value decomposition

Several key functions related to linear algebra are listed in Table 2.4. There are many others. See section 7.1 for tips on using some of these functions.

2.4   Array operations

Array operations act identically on each element of an array. For arrays of identical size, the operations + and - are the same as for matrices. When the operands have different sizes, the operation typically returns an error, with one major exception: a scalar is silently "expanded" to match the size of a matching array operand:

```
>> b+2
ans =
       2
       3
       1

>> A-3
ans =
      -2      -1       0
       1       2       3
       4       5       6
```

Thus, the latter case is *not* interpreted in the mathematical matrix sense of $A - 3I$ for the identity matrix I.

The operators *,',^, and / have matrix interpretations. To get elementwise behavior appropriate for an array, precede the operator with a dot:

```
>> A = [1 2 3; 4 5 6; 7 8 9];
>> C = [1 3 -1; 2 4 0; 6 0 1];
>> A.*C                         % array multiplication
ans =
       1       6      -3
       8      20       0
      42       0       9
```

```
>> A*C                          % matrix multiplication
ans =
    23    11     2
    50    32     2
    77    53     2

>> A./A
ans =
     1     1     1
     1     1     1
     1     1     1

>> 1./A
ans =
    1.0000      0.5000      0.3333
    0.2500      0.2000      0.1667
    0.1429      0.1250      0.1111

>> (C+1i)'
ans =
    1.0000 - 1.0000i  2.0000 - 1.0000i  6.0000 - 1.0000i
    3.0000 - 1.0000i  4.0000 - 1.0000i       0 - 1.0000i
   -1.0000 - 1.0000i       0 - 1.0000i  1.0000 - 1.0000i

>> (C+1i).'
ans =
    1.0000 + 1.0000i  2.0000 + 1.0000i  6.0000 + 1.0000i
    3.0000 + 1.0000i  4.0000 + 1.0000i       0 + 1.0000i
   -1.0000 + 1.0000i       0 + 1.0000i  1.0000 + 1.0000i
```

In addition, most elementary functions, such as sin, exp, etc., act elementwise:

```
>> cos(pi*C)
ans =
    -1    -1    -1
     1     1     1
     1     1    -1

>> exp(C)
ans =
    2.7183     20.0855      0.3679
    7.3891     54.5982      1.0000
  403.4288      1.0000      2.7183
```

Note that in this last case the matrix exponential function $e^C = I + C + C^2/2! + C^3/3! + \cdots$, defined for square matrices, is a completely different animal! For it, use expm, not exp.

Elementwise operators are often useful in functional expressions. Consider evaluating a Taylor polynomial approximation to $\sin(t)$:

```
>> t = (0:0.25:1)*pi/2
t =
        0      0.3927      0.7854      1.1781      1.5708

>> t - t.^3/6 + t.^5/120
ans =
        0      0.3827      0.7071      0.9245      1.0045
```

This is easier and clearer than writing a loop for the calculation. (See section 6.2.) However, because polynomials are so common, they have special status and commands of their own. A polynomial is represented by a vector of its coefficients in decreasing degree order, and it is best evaluated using `polyval`:

```
>> polyval( [1/120,0,-1/6,0,1,0], t )
ans =
        0      0.3827      0.7071      0.9245      1.0045
```

The final zero in the coefficient vector is very important here, since leaving it out would shift all of the coefficients down in degree by one. Leading zeros, on the other hand, have no effect.

Occasionally it is useful to do logical elementwise operations on arrays. The |, &, and ~ operators perform logical elementwise OR, AND, and NOT, respectively. Note, however, that a different form of OR and AND may be preferred in some statements, as explained in section 3.3.

```
>> (t>0) & (t<1)
ans =
      0      1      1      0      0
```

Another kind of array operation works in parallel along one dimension of the array, returning a result that is one dimension smaller:

```
>> C = [1 3 -1; 2 4 0; 6 0 1];

>> sum(C,1)
ans =
      9      7      0

>> sum(C,2)
ans =
      3
      6
      7
```

Other functions that behave this way are shown in Table 2.5.

Table 2.5. *Dimension-reducing functions.*

max	sum	mean	any
min	diff	median	all
sort	prod	std	

2.5   Sparse matrices

It's natural to think of a matrix as a rectangular table of numbers. However, many real-world matrices are both extremely large and very **sparse**, meaning that most entries are zero.[4] In such cases it's wasteful, or just impossible, to store every entry. Instead, one should take advantage of sparsity by storing only the nonzero entries and their locations. MATLAB has a `sparse` data type for this purpose. The `sparse` and `full` commands convert back and forth between the two available internal representations:

```
>> A = vander(1:3);
>> sparse(A)
ans =
   (1,1)        1
   (2,1)        4
   (3,1)        9
   (1,2)        1
   (2,2)        2
   (3,2)        3
   (1,3)        1
   (2,3)        1
   (3,3)        1

>> full(ans)
ans =

     1     1     1
     4     2     1
     9     3     1
```

Sparsifying a standard full matrix is usually not the right way to create a sparse matrix—you should avoid creating very large full matrices, even temporarily. One alternative is to give `sparse` the raw data required by the format. (This is the functional inverse of the `find` command.)

---

[4]For instance, the link structure of the Web can be described by an adjacency matrix in which a_{ij} is nonzero if page j links to page i. Obviously, any page links to a negligible fraction of all Web pages!

```
>> sparse(1:4,8:-2:2,[2 3 5 7])
ans =
    (4,2)          7
    (3,4)          5
    (2,6)          3
    (1,8)          2
```

Alternatively, you can create an empty sparse matrix with space to hold a specified number of nonzeros, and then fill it in using standard subscript assignments.
Another useful sparse building command is spdiags, which builds along the
diagonals of the matrix:

```
>> M = ones(6,1)*[-20 Inf 10]
M =
    -20    Inf    10
    -20    Inf    10
    -20    Inf    10
    -20    Inf    10
    -20    Inf    10
    -20    Inf    10

>> full( spdiags( M,[-2 0 1],6,6 ) )
ans =
    Inf     10      0      0      0      0
      0    Inf     10      0      0      0
    -20      0    Inf     10      0      0
      0    -20      0    Inf     10      0
      0      0    -20      0    Inf     10
      0      0      0    -20      0    Inf
```

The nnz command tells how many nonzeros are in a given sparse matrix.
Since it's impractical to view directly all the entries (even just the nonzeros) of a
realistically sized sparse matrix, the spy command helps by producing a plot in
which the locations of nonzeros are shown. For instance, spy(bucky) shows
the pattern of bonds among the 60 carbon atoms in a buckyball.

MATLAB has a lot of built-in ability to work intelligently with sparse matrices. The arithmetic operators +, -, *, and ^ use sparse-aware algorithms and
produce sparse results when applied to sparse inputs. The backslash \ uses sparse-
appropriate matrix algorithms automatically as well. There are also functions for
the iterative solution of linear equations, eigenvalues, and singular values that
exploit sparsity well. See section 7.2.

Exercises

2.1. Let A be a random matrix generated by rand(8). Find the maximum values (a) in each column, (b) in each row, and (c) overall. Also (d) use find to find the row and column indices of all elements that are larger than 0.25.

2.2. A *magic square* is an $n \times n$ matrix in which each integer $1, 2, \ldots, n^2$ appears once and for which all the row, column, and diagonal sums are identical. MATLAB has a command magic that returns magic squares. Check its output at a few sizes and use MATLAB to verify the summation property. (The antidiagonal sum will be the trickiest. Look for help on how to "flip" a matrix.)

2.3. Are the following true or false? Assume A is a generic $n \times n$ matrix.

   (a) A^(-1) equals 1/A

   (b) A.^(-1) equals 1./A

2.4. Suppose p is a row vector of polynomial coefficients. What does this line do?

$$(\text{length}(p)-1:-1:0) \ .* \ p$$

2.5. (a) Look up diag in the online help and use it (more than once) to build the 16×16 matrix

$$D = \begin{bmatrix} -2 & 1 & 0 & 0 & \cdots & 0 & 1 \\ 1 & -2 & 1 & 0 & \cdots & 0 & 0 \\ 0 & 1 & -2 & 1 & 0 & \cdots & 0 \\ \vdots & \ddots & \ddots & \ddots & \ddots & \ddots & \vdots \\ 0 & \cdots & 0 & 1 & -2 & 1 & 0 \\ 0 & 0 & \cdots & 0 & 1 & -2 & 1 \\ 1 & 0 & 0 & \cdots & 0 & 1 & -2 \end{bmatrix}.$$

   (b) Now read about toeplitz and use it to build D.

   (c) Use toeplitz and whatever else you need to build

$$\begin{bmatrix} 1 & 2 & 3 & \cdots & 8 \\ 0 & 1 & 2 & \cdots & 7 \\ & \ddots & \ddots & & \vdots \\ 0 & 0 & \cdots & 1 & 2 \\ 0 & 0 & \cdots & 0 & 1 \end{bmatrix} \quad \text{and} \quad \begin{bmatrix} 1 & \frac{1}{2} & \frac{1}{3} & \cdots & \frac{1}{8} \\ \frac{1}{2} & 1 & \frac{1}{2} & \cdots & \frac{1}{7} \\ \vdots & \ddots & \ddots & \ddots & \vdots \\ \frac{1}{7} & \frac{1}{6} & \ddots & 1 & \frac{1}{2} \\ \frac{1}{8} & \frac{1}{7} & \cdots & \frac{1}{2} & 1 \end{bmatrix}.$$

   The second case looks best in format rat.

2.6. Suppose A is any matrix. What does this statement do?

```
A( 1:size(A,1)+1:end )
```

2.7. (a) Suppose A is a matrix whose entries are all positive numbers. Write one line that will multiply each column of A by a scalar so that, in the resulting matrix, every column sums to 1.

   (b) Try this more difficult variation: Suppose that A may have zero entries, and leave a column of A that sums to zero unchanged.

2.8. Find a MATLAB one-line expression to create the $n \times n$ matrix A satisfying

$$a_{ij} = \begin{cases} 1 & \text{if } i - j \text{ is prime,} \\ 0 & \text{otherwise.} \end{cases}$$

2.9. Suppose we represent a standard deck of playing cards by a vector v containing one copy of each integer from 1 to 52. Show how to "shuffle" v by rearranging its contents in random order. (Note: One very easy answer to this problem can be found if you look hard enough.)

2.10. Let B=bucky, and make a series of spy plots of B^2, B^3, etc. to see the phenomenon of *fill-in*: many operations, including multiplication, increase the density of nonzeros. Can you see why the (i, j) entry of B^n is the number of paths of length n between nodes i and j? What fill-in do you see with inv(B)?

Chapter 3

Scripts and Functions

An **M-file** is a plain text file containing MATLAB commands and saved with the filename extension .m. There are two types, **scripts** and **functions**. MATLAB comes with a good editor that is tightly integrated into the environment; start it by typing `edit`. However, you are free to use any text editor. An M-file should be saved in the path in order to be executed. Recall that the path is a list of directories (folders) in which MATLAB will look for files. Use `editpath` or menus to see and change the path.

There is no need to compile either type of M-file. Simply type in the name of the file (without the extension) in order to run it. Changes that are saved to disk will be included in the next call to the function or script.

An important type of statement in any M-file is a **comment**, which is indicated by a percent sign `%`. Any text on the same line after a percent sign is ignored, unless `%` appears as part of a string. Furthermore, the first contiguous block of comments in an M-file serves as documentation for the file and will be typed out in the command window if `help` is used on the file. For instance, say the following is saved as myscript.m on the path:

```
% This script solves the nasty homework problem
% assigned by Professor Driscoll.

x = rand(1);  % He'll never notice.
```

Then at the prompt one would could obtain

```
>> help myscript

  This script solves the nasty homework problem
  assigned by Professor Driscoll.
```

Normally, you will want to include syntax information as well as a short description within the documentation of your own functions.

3.1   Using scripts effectively

The contents of a script file are literally interpreted as though they were typed at the prompt. In fact, you might prefer to use MATLAB exclusively by typing all commands into scripts and running them. This technique makes it easier to create and revise a sequence of more than a few lines, and it helps you document and retrace your steps later. Note too that you can create a script from lines you have already entered by highlighting them in the Command History window of the desktop and right-clicking on them.

Another interesting use for scripts is to publish them as HTML files (Web pages) or in other available formats. This is a nice way to create reports based on MATLAB results, although support for mathematical expressions is currently rather limited. When a script is published, blocks of commented lines appear as regular text, and code and resulting output are set off in nice formats. To get the best results, use double percent signs at the beginning of a line to create a *cell boundary*. These boundaries cause output to be created before the next block of comments, code, and output begins.

For example, the script

```
%%
% A Hilbert matrix has entries that are reciprocals of
% integers.
format rat
H = hilb(4)

%%
% Although the inverse of a Hilbert matrix is difficult
% to find numerically by Gaussian elimination, an explicit
% formula is known.
invhilb(4)

%%
format short e
H*ans
```

will publish to look something like this:

A Hilbert matrix has entries that are reciprocals of integers.

```
format rat
H = hilb(4)

H =
```

1	1/2	1/3	1/4
1/2	1/3	1/4	1/5
1/3	1/4	1/5	1/6
1/4	1/5	1/6	1/7

Although the inverse of a Hilbert matrix is difficult to find numerically by Gaussian elimination, an explicit formula is known.

```
invhilb(4)
```

```
ans =

         16         -120          240         -140
       -120         1200        -2700         1680
        240        -2700         6480        -4200
       -140         1680        -4200         2800
```

```
format short e
H*ans
```

```
ans =

          0            0            0            0
          0            0            0            0
          0            0            0  -5.6843e-014
          0            0            0            0
```

In addition, cells allow you to create titled sections, complete with a hyperlinked table of contents. The built-in Editor is the best way to publish scripts. It will help you to create and work with cells and to clearly show cell boundaries and titles, and it lets you publish and view results with minimal fuss.

As a rule of thumb, call scripts only from the command line, and do not call other scripts from within a script. For automating individual tasks that fit within a larger framework, functions are the better choice, as explained in the next section.

3.2 Functions and workspaces

Both functions and scripts gather MATLAB statements to perform complex tasks. The most important distinguishing feature of a function is its **local workspace**. Any variables created while the function executes are available only within that invocation of the function, unless you go out of your way to bend the rules (see section 6.4). Conversely, the variables available to the command line—those in the **base workspace**—are normally not visible within the function. If other functions are called during a function's execution, each of those secondary calls also sets up a private local workspace. These restrictions on data access are called **scoping**, and they make it possible for you to write complex programs that use many components without worrying about variable name clashes. You can always see variables in the current workspace by typing `whos`, or in the Workspace window of the desktop. At the command line, the base workspace is ordinarily in context, but during function debugging you can inspect other local workspaces.

Each function starts with a line such as

```
function  [out1,out2]  =  myfun(in1,in2,in3)
```

The name `myfun` should match the name of the file on disk. The variables `in1`, etc. are **input arguments**, and `out1`, etc. are **output arguments**. You can have as many as you like of each type (including zero) and call them whatever you want. Theoretically, the only communication between a function's workspace and that of its caller is through the input and output arguments, though there are some exceptions (see section 6.4). The values of the input arguments to a function are *copies* of the original data, so any changes you make to them will not change anything outside of the function's scope.[5]

Here is a function that implements the quadratic formula for finding the roots of $ax^2 + bx + c$:

```
function  [x1,x2]  =  quadform(a,b,c)

d  =  sqrt(b^2  -  4*a*c);
x1  =  (-b  +  d)  /  (2*a);
x2  =  (-b  -  d)  /  (2*a);
```

(It must be pointed out that from a numerical standpoint, this is not a good algorithm; see Exercise 3.1.) With this text saved as quadform.m in the MATLAB path, you could then immediately enter

```
>>  [r1,r2]  =  quadform(1,1,1)

r1  =
   -0.5000  +  0.8660i

r2  =
   -0.5000  -  0.8660i
```

The main use of a function is to compartmentalize a specific task. Any complex problem is decomposed into a series of smaller steps, and the scoping rules of functions allow you to deal with each step independently. They also let you exploit well-crafted solutions to fundamental tasks that appear over and over in different problems. Like a good theorem, a good function invites you to inspect the details of its construction once and then forget about them.

---

[5]Actually, the MATLAB interpreter avoids copying a function argument—that is, it "passes by reference" using a pointer—if the function never alters the value of that argument. Furthermore, in MATLAB 7.6 and later versions, you can force passing by reference even for modified values; see the help on `handle` for details.

Another important aspect of function M-files is that many of the functions that come with MATLAB are themselves M-files that you can read and borrow from. This is an excellent way to learn good programming practice—and dirty tricks.

3.3 Conditionals: `if` and `switch`

Often, a function needs to branch based on run-time conditions. MATLAB offers features for this that are similar to those in most languages.

Here is an example illustrating most of the features of `if`.

```
if isinf(x) || ~isreal(x)
  disp('Bad input!')
  y = NaN;
elseif (x == round(x)) && (x > 0)
  y = prod(1:x-1);
else
  y = gamma(x);
end
```

The conditions for `if` statements may involve the relational operators of Table 2.3 on page 17, or functions such as `isinf` that return logical values. Numerical values can also be used, with nonzero meaning true, but "`if x~=0`" is better form than "`if x`" when x is numeric. Take some care when using arrays to construct the condition of an `if` statement. When the condition is not a scalar, it is taken as true only when *all* of the elements are true/nonzero. To avoid confusion, it's best to use `any` or `all` to reduce logical arrays to scalar values.

Compound scalar conditions should be created using the operators `&&` (logical AND) and `||` (logical OR), rather than the array operators `&` and `|`. (The NOT operator `~` is the same in both cases.) The double-symbol operators can be **short-circuited**: if, as a condition is evaluated from left to right, it becomes obvious before the end that truth or falsity is assured, then evaluation of the condition is halted. This feature makes it convenient to write things like

```
if (length(x) > 2) && (x(3)==0), disp('xy plane'), end
```

that otherwise could create errors or be awkward to write.

In some situations, the `isequal` command is more robust than the `==` relational operator. For instance, the syntax `isequal(s,'foo')` returns false if s is not identical to the string `'foo'`, whereas `s=='foo'` would throw an error if s were not of size 1×3. Usually, the former behavior is more desirable.

The `if/elseif` construct is preferred when only a small number of alternatives are present. When a large number of options are needed, it's customary to use instead. For instance:

```
switch units
  case 'length'
    disp('meters')
  case 'volume'
    disp('liters')
  case 'time'
    disp('seconds')
  otherwise
    disp('I give up')
end
```

The switch expression can be a string or a number. The first matching case has its commands executed.[6] If otherwise is present, it gives a default option if no case matches.switch

3.4 Loops: for and while

Many algorithms require **iteration**, the repetitive execution of a block of statements. Again, MATLAB is similar to other languages in this area. It's worth repeating here that because i and j are very tempting as loop indices, you are advised to always use 1i or 1j as the imaginary unit $\sqrt{-1}$.

As an illustration of a simple for loop, consider this code for computing 10 members of the famous Fibonacci sequence:

```
>> f = [1 1];
>> for n = 3:10
     f(n) = f(n-1) + f(n-2);
   end
```

You can have as many statements as you like in the body of a loop. In this example, the value of the index n will change from 3 to 10, with an execution of the body after each assignment. Remember that 3:10 is really just a row vector. In fact, you can use *any* row vector in a for loop, not just one created by a colon. For example,

```
>> x = 1:100;  s = 0;
>> for j = find(isprime(x))
     s = s + x(j);
   end
```

This finds the sum of all primes less than 100. For a better version, though, see page 64.

---

[6]Execution does not "fall through" as in C.

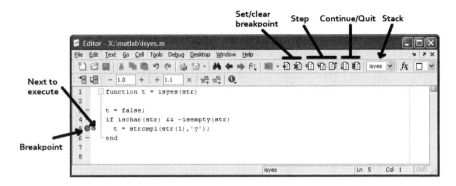

Figure 3.1. *Debugging tools in the MATLAB Editor.*

It is sometimes necessary to repeat statements based on satisfying a condition rather than a fixed number of times. This is done with `while`:

```
while abs(x) > 1
  x = x/2;
end
```

The condition is evaluated before the body is executed, so it is possible to get zero iterations. It's often a good idea to limit the number of repetitions to avoid infinite loops (as could happen above if `x` is infinite). This can be done in a number of ways, but the most common is to use `break`.

```
n = 0;
while abs(x) > 1
  x = x/2;
  n = n+1;
  if n > 50, break, end
end
```

A `break` immediately jumps execution to the first statement after the loop. It's good practice to include some diagnostic output or other indication that an abnormal loop exit is taking place.

3.5 Debugging and profiling

MATLAB provides a debugging mode that lets you pause execution anywhere inside an M-file function. It's a great way to fix faulty code or to understand how a code works. A screenshot of the debugging tools is shown in Figure 3.1.

To enter debugging mode, you set **breakpoints** in one or more functions. In the Editor, you click on the dash next to a line number to do this, or you can use the Debug menu to create conditional breakpoints. When MATLAB stops at a breakpoint, you can inspect and modify all the variables currently in scope—in fact, you can execute anything at all from the command line. You can then continue function execution normally or step by step. See `help debug` for all the details.

Another common need is to make working code run faster. The most efficient way to do so is to locate which parts are taking the most time to execute. These lines are then obvious candidates for optimization. You can find such lines by **profiling**, which keeps track of time spent on every line of every function. Profiling is also one way to determine function dependencies (who calls whom). Get started by typing `profile viewer`.

Exercises

3.1. Write a function `quadform2` that implements the quadratic formula differently from `quadform` above (page 30). First compute

$$x_1 = \frac{-b - \text{sign}(b)\sqrt{b^2 - 4ac}}{2a},$$

which is the root of largest magnitude, and then use the identity $x_1 x_2 = c/a$ to find x_2. Apply both `quadform` and `quadform2` to find the roots of $x^2 - (10^7 + 10^{-7})x + 1$. Do you see why `quadform2` is better?

3.2. The degree-n Chebyshev polynomial is defined by

$$T_n(x) = \cos\left[n\cos^{-1}(x)\right], \quad -1 \le x \le 1.$$

These satisfy $T_0(x) = 1$, $T_1(x) = x$, and the recursion relation

$$T_{n+1}(x) = 2xT_n(x) - T_{n-1}(x), \quad n \ge 1.$$

Write a function `chebeval(x,N)` that evaluates all of the Chebyshev polynomials of degree less than or equal to N at all of the points in column vector x. The result should be an array of size `length(x)` by N+1.

3.3. One way to compute the exponential function e^x is to truncate its Taylor series expansion around $x = 0$,

$$e^x = 1 + x + \frac{1}{2!}x^2 + \frac{1}{3!}x^3 + \cdots.$$

Unfortunately, many terms are required for accuracy if $|x|$ is large. But a special property of the exponential is that $e^{2x} = (e^x)^2$. This leads to a *scaling and squaring* method: Divide x by 2 repeatedly until $|x| < 1/2$, use the Taylor series (16 terms should be more than enough), and square the result repeatedly. Write a function `expss(x)` that performs these three steps. (The functions `cumprod` and `polyval` can help with evaluating the Taylor expansion.) Test your function on x values $-30, -3, 3, 30$.

3.4. Let x and y be column vectors describing the vertices of a polygon, given in order. Write functions `polyperim(x,y)` and `polyarea(x,y)` that compute the perimeter and area of the polygon. For the area, use a formula based on Green's theorem:

$$A = \frac{1}{2}\left|\sum_{k=1}^{n} x_k y_{k+1} - x_{k+1} y_k\right|.$$

Here n is the number of polygon vertices, and by definition, $x_{n+1} = x_1$ and $y_{n+1} = y_1$. Test your functions on a square and an equilateral triangle.

3.5. Suppose a data source produces a series of characters drawn from a set of M distinct symbols. If symbol k is produced with probability p_k, the *first-order entropy* of the source is defined as

$$H_1 = -\sum_{k=1}^{M} p_k \log_2 p_k.$$

Essentially H_1 is the number of bits needed per symbol to encode a long message; that is, it measures the amount of information content, and therefore the potential success of compression strategies. The value $H_1 = 0$ corresponds to the case of only one symbol being produced—no information—while if all M symbols have equal probability, then $H_1 = \log_2 M$.

Write a function `[H,M] = entropy(v)` that computes entropy for a vector v. The probabilities should be computed empirically by finding the unique entries (using `unique`), then counting the occurrences of each symbol and dividing by the length of v. Try your function on some built-in image data by entering `load clown, v = X(:);`.

Chapter 4

More on Functions

The previous chapter covered the bare minimum on MATLAB functions. Here are some more advanced concepts and tools that at some point you will probably find very useful.

4.1 Function handles and anonymous functions

One way to look at a function is as an algorithm or process. Sometimes, though, we might prefer to look at a function as data. For instance, the mathematical problem of definite integration could be seen as a map from inputs in the form of an interval $[a,b]$ and a function $f(x)$ to the value of $\int_a^b f(x)\,dx$. As you might expect, this is a classical problem in computing, and there are many algorithms available to approximate this map. Thus we can find ourselves in the situation of writing or using a function that accepts another function as data. MATLAB likes to call these **function functions**, and you can get a list of them by typing `help funfun`. Other common problems that involve operation on functions as data are finding roots, optimization, and solving differential equations.

MATLAB requires you to use a special construct called a **function handle** to distinguish between the invocation of a function and the abstraction of that function as data. For example, the built-in `fzero` attempts to find a root of a function supplied as its first input argument, starting from a guess in the second argument. We cannot call it to find a root of $\sin(x)$ by typing

```
>> fzero(sin,3)   % error!
```

because the MATLAB interpreter would first try to call the built-in `sin` function with no inputs, which is an error. Instead, you must create a function handle by using the @ prefix, as in

```
>> fzero(@sin,3)
ans =
    3.1416
```

Any callable function in scope can be converted into a handle, including built-ins and your own M-file functions.

Often, you need a function definition that combines built-in operations in an elementary way. The best tool for this situation is an **anonymous function**. An anonymous function lets you simultaneously define a function and create a handle to it.[7] For instance, to create a representation of the mathematical function $\sin(x) + \cos(x)$, you would enter

```
>> sincos = @(x) sin(x) + cos(x);
```

Now sincos is a function of one argument, and the syntaxes sincos(pi/4) and sincos(rand(2)) are perfectly acceptable. It is also straightforward to create an anonymous function of multiple arguments, as in

```
>> w = @(x,t,c) cos(x-c*t);
```

There is no need to assign the anonymous function to a named variable, however, as in this syntax:

```
>> fzero( @(x) sin(x)+cos(x), 0 )
ans =
    -0.7854
```

When you define an anonymous function, the variable names following the @ symbol are inputs to the function. If the same names are defined in the current workspace, those values are unavailable within the anonymous definition. However, other values defined in the current workspace at the time of definition *are* available to the anonymous function definition. Furthermore, any such values used within the definition are "baked in" to the anonymous function: they remain fixed, even if the values are later changed or the defining workspace goes out of scope.

This feature can be amazingly convenient. A prototypical situation is that you have an M-file function bigfun(a,x,b) that accepts parameters a and b in addition to a variable x of interest. Syntactically, many function functions accept only a function of one variable as data. Hence you define an anonymous function wrapper such as

```
>> f = @(x) bigfun(0.5,x,b_value);
```

that locks in particular parameter values and accepts only one input variable. For example, consider again fzero, which finds a root of a function of one variable. Suppose you wish to find a solution of $e^{-ax} = x$ for multiple values of the parameter a. An anonymous function makes this a snap:

---

[7]Before anonymous functions became available, **inline functions** offered some of the same capability in older versions of MATLAB. Inline functions are less useful in several ways, however, and offer no real advantages.

```
>> for a = 1:0.25:2, fzero( @(x) exp(-a*x)-x, 0 ), end
ans =
    0.5671
ans =
    0.5212
ans =
    0.4839
ans =
    0.4528
ans =
    0.4263
```

In the body of the loop, the variable a is assigned each of the values 1, 1.25, 1.5, 1.75, 2 in turn. Each time, a new anonymous function is created to lock in that value of a and create a function of x alone, which is fed to fzero.

4.2   Subfunctions and nested functions

A single M-file may hold more than one function definition. The function header line at the top of a file defines the **primary function** of the file. Two other types of functions can be in the same file: subfunctions and nested functions.

A **subfunction** is mostly a convenient way to avoid directory and name clutter. The subfunction begins after the end of the primary function, with a new function header line. Every subfunction in the file is available to be called by the primary function and the other subfunctions. In all respects, it behaves like a primary function in a separate file, including the use of a private variable workspace. As a silly example of a function using a subfunction, consider a new version of quadform.m:

```
function [x1,x2] = quadform(a,b,c)
  d = discrim(a,b,c);
  x1 = (-b + d) / (2*a);
  x2 = (-b - d) / (2*a);
end   % quadform()

function D = discrim(A,B,C)
  D = sqrt(B^2 - 4*A*C);
end   % discrim()
```

The end line is optional for single-function files, but it is a good idea when subfunctions are involved, and it is mandatory when using nested functions. Changes made to a, b, or c inside discrim would not propagate into the rest of quadform, nor would any other variables in the primary function be available to discrim.

Outside of the defining file, subfunctions are not visible—only the primary function can be called. However, it is possible to exploit a major and useful exception to this rule. Suppose you want to investigate the eigenvalues of a certain matrix $A(x)$—specifically, you wish to find the smallest x such that the maximum real part of the eigenvalues equals 1. Here is a simple approach using `fzero` to reach the target value:

```
function x0 = findx
  x0 = fzero(@objective,[0 10]);
end

function r = objective(x)
  B = diag(ones(499,1),1);   A = B-B';
  A(1,1) = x;
  e = eig(A);   r = max(real(e)) - 1;
end
```

The subfunction `objective` is passed as a handle to `fzero`, which is able to use the handle to access and call the subfunction, even though it would be out of scope if `fzero` were to try to invoke it directly by name.

   A **nested function** is similar to a subfunction, but it is defined *within* the scope of a parent function. It behaves differently from a subfunction in one very important respect: its variable workspace can overlap that of its parent. Any variable used in both the nested function and its parent is shared.

   One reason for using a nested function is to create lasting side effects. Consider again the eigenvalue example just above. Because `fzero` is a "black box" with a standardized syntax, it can return only the value of x we seek. It cannot tell us, for example, the imaginary part of the eigenvalue whose real part equals 1, nor can it tell us any other information about the eigenvalues. To get that information, we would have to repeat much of the `objective` function once again, which makes for messy code and extra computation time. Reorganizing the computation using a nested function, however, allows us to pass extra information out of the objective calculation:

```
function x0 = findx
  function r = objective(x)
    A(1,1) = x;
    e = eig(A);   r = max(real(e)) - 1;
  end

  B = diag(ones(499,1),1);   A = B-B';
  e = [];   % create a shared variable for side effect
  x0 = fzero(@objective,[0 10]);
  plot(e,'x')
end
```

The line e = [] ; is critical. Giving the variable e a definition in the parent workspace allows it to be shared between the nested objective and its parent, so that at the termination of fzero, the value of e is the same as what it was at the end of the most recent eigenvalue computation. Notice that we also passed static information about the variable A *into* the nested function, potentially saving otherwise wasted time over the previous version.

Variable sharing with nested functions is a technique best used cautiously. Experience shows that reliance on common workspaces can lead to code that is easily broken and confusing to read. There are situations like this one, however, in which the technique offers possibilities that are hard to duplicate otherwise.

4.3   Errors and warnings

MATLAB functions may encounter statements that are impossible to execute, in a way that could not have been predicted until the execution is to take place. For example, the statement A*B is syntactically valid but meaningless in context when A and B are defined as matrices of incompatible sizes. In such a situation, an **error** is thrown: execution halts, a message is displayed, and control is returned to the prompt, with the output arguments of the function ignored. You can throw errors in your own functions with the error statement, called with a string that is displayed as the message. Similar to an error is a **warning**, which displays a message but allows execution to continue. You create such messages using warning.

Sometimes you would like the ability to recover from an error in a subroutine and continue with a contingency plan. This can be done using the try/catch construct. For example, the following will continue asking for a statement until you give it one that executes successfully:

```
done = false;
while ~done
  state = input('Enter a valid statement: ','s');
  try
    eval(state);
    done = true;
  catch me
    disp('That was not a valid statement! Look:')
    disp(me.message)
  end
end
```

Within the catch block you can find the most recent error message using lasterr, or (as is preferred in recent MATLAB versions) by inspecting the exception object me, as shown above and explained in the help pages for MException.

4.4   Input and output arguments, revisited

Good programming practice dictates that you check the input arguments to a function for validity. This is an especially good idea in a weakly typed language such as MATLAB, where valid syntax can lead to meaningless results. There are a number of helper functions for checking inputs, as demonstrated in the following fragment:

```
function [x,y] = myfun(a,b,c)

assert(isnumeric(a), 'First input must be a number.')
assert(numel(a)==1, 'First input must be a scalar.')
assert(~any(isinf(b)), 'Second input must be finite.')
assert(~any(isnan(b)), 'No NaNs allowed in second input.')
assert(ischar(c), 'Third input must be a string.')
```

If the first expression given to `assert` is not true, an error is thrown with the message and is given as the second argument.

Although a function's header spells out the number and names of the input and output arguments, these are not enforced at the time of function invocation. Instead, execution proceeds as long as possible until something undefined or otherwise illegal happens. This fact allows you to accept different numbers of input arguments in different situations. One common use of variable-length argument lists is to give default values to missing inputs, as in the following fragment:

```
function shirt(neck,sleeve,color,cuff)

if nargin < 4
  cuff = 'button';
  if nargin < 3
    color = 'blue';
  end
end
```

The value of `nargin` is always set at run-time to the number of input arguments with which the function was actually called. In the syntax above, default values are assigned to the third and fourth arguments if they are missing. (The inputs are place-sensitive, however, so there is no way to specify a `cuff` without also specifying a `color` in this setup.) If the function is called with just zero or one inputs, execution will continue unless and until a code reference is made to `neck` or `sleeve`, which are not given defaults.

For a different take on allowing flexibility in the number of input arguments, see section 6.6.

Exercises

4.1. Write a function `plusone(f,x)` that, given a function f and value x, returns $f(x)+1$.

4.2. Write a function `trap(f,a,b,n)` that computes the trapezoidal rule approximation to $\int_a^b f(x)\,dx$,

$$\frac{h}{2}\big[f(x_0)+2f(x_1)+2f(x_2)+\cdots+2f(x_{n-1})+f(x_n)\big],$$

where $h = (b-a)/n$ and $x_i = a+ih$. Test your function on $\sin(x)+\cos(x)$ for $0 \le x \le \pi/3$. For a greater challenge, write a function `simp` for Simpson's rule,

$$\int_a^b f(x)\,dx \approx \frac{h}{3}\big[f(x_0)+4f(x_1)+2f(x_2)+\cdots+4f(x_{n-1})+f(x_n)\big].$$

This formula requires n to be even. You may choose to check the input for this.

4.3. Write a function `bisect(f,a,b,tol)` that employs the *bisection* method for finding a value of x such that $f(x) = 0$. The first input is a handle to a function computing f. Assuming that f is continuous and that $f(a)f(b) < 0$, the function has at least one root in the interval (a,b). Define $m = (a+b)/2$, and if $f(m) \ne 0$, then either $f(a)f(m) < 0$ or $f(b)f(m) < 0$, so the root is either in (a,m) or (m,b). Continue the process until the root is contained in an interval of length less than `2*tol`, and stop. The best code takes care not to evaluate the given `f` more often than necessary.

4.4. Write a function `newton(f,fprime,x0,tol)` that implements Newton's iteration for rootfinding on a scalar function:

$$x_{n+1} = x_n - \frac{f(x_n)}{f'(x_n)}.$$

The first two inputs are handles to functions computing f and f', and the third input is an initial root estimate. Continue the iteration until either $|f(x_{n+1})|$ or $|x_{n+1} - x_n|$ is less than `tol`.

4.5. Modify `newton` from the previous exercise so that it works on a system of equations $\mathbf{F}(\mathbf{x})$. The function `fprime` now returns the Jacobian matrix J, and the Newton update is written mathematically as

$$\mathbf{x}_{n+1} = \mathbf{x}_n - \mathsf{J}^{-1}\mathbf{F}(\mathbf{x}_n),$$

although in numerical practice one does not compute the inverse of the Jacobian but solves a linear system of equations in which $\mathbf{F}(\mathbf{x}_n)$ is the right-hand side.

4.6. Many simple financial instruments that have regular equal payments, such as car loans or investment annuities, can be modeled by the equation

$$F = P\left(\frac{(1+r)^t - 1}{r}\right),$$

where P is the regular payment, r is a fixed interest rate (say, $r = 0.05$ for 5% interest), t is the number of payment intervals elapsed, and $F(t)$ is the accumulated value of the instrument at time t. This equation is not easily solved for r.

Write a script or function that finds r when $P = 200$, $t = 30$, and F takes the values $10000, 15000, \ldots, 40000$.

Chapter 5
Graphics

Graphical display is one of the greatest strengths of the MATLAB software, and one of its most complicated subjects. The basics are quite simple, but you also get complete control over practically every aspect of each graph, and with that power comes some complexity.

Graphical objects are classified by **type**, the most important of which are shown in Figure 5.1. The available types lie in a strict hierarchy: each data-relevant object, such as a **line** or **surface**, must lie in an **axes**, which in turn lies in a **figure** window, descended from a virtual **root**. Of the command names in Figure 5.1, the most useful is `figure`, which by itself opens a new figure window. The others, such as `line` and `surface`, are considered low-level functions and are not usually called directly. Instead you use friendlier functions that create these object types inside the most recently created or clicked-on axes. Similarly, a **group** or **series** graphics object is typically a collection of more primitive objects that are created and bundled together by a convenient routine.

Each rendered graphics object has properties that you can set at creation time or after the fact to control its appearance. You can also add axes labels, titles, and other annotations to the graph. Most graphical modifications can be done either via the command line or by using the GUI (graphical user interface). In particular, the Edit Plot button, which has the icon ⬉ , enables you to double-click or right-click virtually any graphical object to change its attributes. While there are examples of making changes in which one method is clearly preferred, for the most part the choice is a matter of personal comfort.

5.1 Data plots versus function plots

Ultimately, any plot comes down to the display of numerical values. The source of those values, however, dictates how you generate or use them. In many

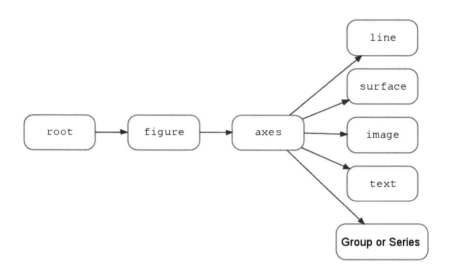

Figure 5.1.   *A partial graphics object hierarchy.   Except for the group/series case, each name is a graphics object type as well as a low-level command name.*

applications, the source is observed data or discrete values of an unknown function, such as the solution of a differential equation. In this case, you conceptually want to display a fixed set of points in the plane or space, possibly with additional visual information such as colors or connecting lines. The "traditional" and better-known plotting functions in MATLAB follow this model.

Sometimes, though, the source of displayed data is one or more explicitly known mathematical functions. Here one would like to work with the functions themselves and keep the possibility of evaluating them as needed to generate the points that make a nice-looking plot. MATLAB has a number of functions, all starting with the letters `ez`, for plots made in this way. The `ez` functions are the ones that work the most like the default plotting functions in *Mathematica*® and Maple® and a symbolic scientific calculator. However, keep in mind that once points are generated and rendered, the original source of them is forgotten and becomes irrelevant—so, for instance, zooming in on part of a plot will not refine the graph, as it does in many symbolic packages.

5.1.1   ez plots

The command `ezplot` makes two-dimensional (2D) plots of explicit, implicit, or parametric functions. These are represented mathematically by $y = f(x)$, $F(x, y) = 0$, or the pair $x = f(t)$, $y = g(t)$, respectively. You must supply `ezplot` with one or more function handles to define the plot, and define a range for the variables, as in these examples:

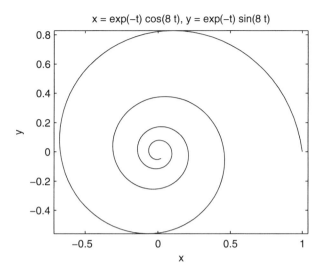

Figure 5.2. *Result of an* ezplot *command.*

```
>> ezplot( @sin,  [0 2*pi] )
>> ezplot( @(x,y) x.^4+y.^4-1,  [-1 1] ), axis equal
>> x= @(t) exp(-t).*cos(8*t);
>> y= @(t) exp(-t).*sin(8*t);
>> ezplot(x,y,[0 3])
```

The results of the last command are shown in Figure 5.2. Note that all of the anonymous function definitions must use array operators such as `.^` and `.*`. The `axis equal` command used in the second example makes the coordinates have a one-to-one aspect ratio, so that circles don't look like ellipses. The `ezpolar` command also makes a curve in the plane, given a function that defines r as a function of θ. The `ezplot3` command plots parametric space curves and can animate "walking" along the curve:

```
>> ezplot3(@cos, @sin, @(t) exp(-t/8), [0 40], 'animate')
```

One way to visualize a function of two variables is to use a contour plot, in which a family of curves $F(x,y) = c$ is plotted in the x-y plane for multiple choices of c. There are two `ez` variants for this:

```
>> ezcontour(@(x,y) 4*x.^2-x.*y+y.^2, [-1 1 -2 2] )
>> ezcontourf(@(x,y) sin(3*x-y).*sin(x+2*y), [-pi pi])
```

The second form fills each intercontour region with a different solid color for a more dramatic plot.

For an explicit surface $z = F(x,y)$, or a parametric surface in which x, y, and z each is a function of parameters (u,v), use `ezsurf`. A wireframe plot is produced if you use `ezmesh`.

```
>> ezsurf( @(x,y) x.^2-y.^2-1, [-1 1 -1 1] )
>> x=@(u,v) cosh(u).*cos(v); y=@(u,v) sinh(u).*cos(v);
>> z=@(u,v) sin(v);  ezmesh( x,y,z, [-1 1 0 2*pi] )
```

In both types of plots, color is used as an additional cue to indicate the value of the z coordinate.

You can find many more details and options in the online documentation than are covered here.

5.1.2 Two-dimensional data plots

For discrete data in two dimensions, use `plot`. For example, MATLAB comes with U.S. census data from 1790–1990:

```
>> load census
>> plot(cdate,pop)
```

The plot simply connects dots at the points defined by the values in the two co-ordinate vectors. You can change the line style graphically after the fact or at the time of creation, as in

```
>> plot(cdate,pop,'ro')   % red circles
>> plot(cdate,pop,'k:')   % black dotted lines
>> plot(cdate,pop,'ys-')  % yellow squares and lines
```

The default is for any new plot to replace the existing one. To add curves to the current axes instead, use `hold on`.

This particular data set looks a bit better when the vertical axis is logarith-mically scaled:

```
>> semilogy(cdate,pop,'p-')
```

Here you can see the data very nearly represents exponential growth at least up to 1860. We can add a crude exponential extrapolation from the 1900 data as follows:

```
>> t = 1900:10:2030;
>> hold on
>> semilogy( t, exp(-20.44+0.01306*t), 'k--' )
```

The result is shown in Figure 5.3. Note that the axes limits that frame the data are chosen automatically to be "nice" numbers. If instead you want to minimize the whitespace, use `axis tight` to make the frame as small as possible.

Other useful 2D plotting commands are given in Table 5.1. See a bunch more by typing `help graph2d`. Also explore the figure window buttons and menus, which should be fairly self-explanatory, and try right-clicking on objects to modify their appearance.

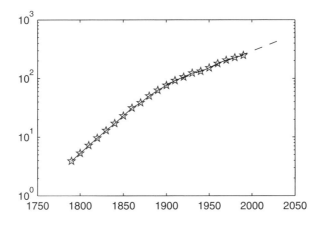

Figure 5.3. *Semilog plot of U.S. census data, and an extrapolation.*

Table 5.1. *2D plotting commands.*

`figure`	Open a new figure window.
`subplot`	Create multiple axes in one figure.
`semilogx, semilogy, loglog`	Scale axes logarithmically.
`axis, xlim, ylim`	Set axes limits.
`legend`	Create legend for multiple curves.
`print`	Send to printer.

5.1.3 Three-dimensional data plots

Plots of surfaces and the like for functions $z = f(x, y)$ operate on the connect-the-dots principle, but the details are more difficult. The first step is to create a grid of points in the x-y plane where f is evaluated to obtain the "dots" in three dimensions. This is most easily done using `meshgrid`.

Here is a typical example:

```
>> x = pi*(0:0.02:1);
>> y = 2*x;
>> [X,Y] = meshgrid(x,y);
```

These arrays define the coordinates of a 2D grid, as can be seen by using

```
>> plot(X,Y,'k.')
```

With this grid definition, we can define some height data in the z direction by array operations:

```
>> Z = sin(X.^2+Y);
>> surf(X,Y,Z)
```

Table 5.2. *3D plotting commands.*

`surf, mesh, waterfall`	Create surfaces in three dimensions.
`colorbar`	Show color scaling.
`plot3`	Plot curves in space.
`pcolor`	Show top view of a colored surface.
`contour, contourf`	Create contour plot.

The array Z is actually an array of values of $f(x,y) = \sin(x^2 + y)$ evaluated on the grid. Finally, `surf` makes a solid-looking surface in which color and apparent height describe the given values of f. An alternative command `mesh` is similar to `surf` but makes a "wireframe" surface. The most common 3D plotting commands are shown in Table 5.2.

Because of mathematical conventions, the output of `meshgrid` is an endless source of confusion. The problem is that for the graph of a function $f(x,y)$, the first variable is interpreted as horizontal and the second variable as vertical. In an array, A, however, the first index dimension is vertical (row) and the second is horizontal (column).[8] The function `meshgrid` and the 3D plotting functions adopt the *graphical* convention. Hence, using the definitions above, the grid point $(x(i),y(j))$ would be found in the arrays above as $(X(j,i),Y(j,i))$. You might choose to avoid this issue by using `ndgrid`, which adopts the array convention, instead of `meshgrid`, but then all calls to `surf` and its ilk need to have transposes on all the arrays, or the graph itself will have x and y transposed.

It's also useful to realize that, while the data must be given at points on a mesh, the mesh has to be *logically* rectangular, not literally rectangular. For instance, you can plot data on the unit disk as follows:

```
>> [R,T] = meshgrid( 0:.02:1, pi*(-1:.05:1) );
>> X = R.*cos(T);   Y = R.*sin(T);
>> pcolor(X,Y,X.^2-Y.^3), axis equal
```

Any surface parameterized by two variables defined on a rectangle can be visualized in the same way.

5.2 Annotation

Graphs of data usually need labels and maybe a title. For example:

```
>> ezplot(@(t) exp(-t/5).*sin(t), [0 6*pi])
>> xlabel('time')
>> ylabel('amplitude')
>> title('Damped oscillator')
```

---

[8]An additional complication, ignored by the plotting routines, is that the vertical direction meaning "increasing" is reversed in the two interpretations. For plots that represent a matrix, use `axis ij` to reverse the vertical axis.

These can also be added and edited by enabling the Edit Plot button and double-clicking the axes. You can also add legends, text, arrows, or text/arrow combinations, to help label data. See the Insert menu on the figure window.

MATLAB supports a limited subset of TEX processing in text strings, including axes labels and titles. This is an easy way to get Greek letters, superscripts, and symbols.

```
>> xlabel('time \tau')
>> legend('{\it e}^{-\tau/5} sin(\tau)')
```

You can get even more elaborate mathematical annotation by requiring a LATEX interpreter for your string, as explained in the documentation. This route can be effective for a stand-alone figure; ironically, though, it may not produce results that blend well with a LATEX document, since such documents are usually customized with respect to fonts and macros. Instead, you might consider the excellent psfrag package for LATEX, which allows you to replace strings in an EPS figure with arbitrary boxes of processed text from within your document.

5.3 Handles and properties

Every rendered object has an identifier or **handle**. The functions `gcf`, `gca`, and `gco` return the handles to the active figure, axes, and object (usually the most recently drawn or clicked-upon). The handle of a figure is always the figure number; other handles have arbitrary values.

Properties can be accessed and changed at the command line by the functions `get` and `set`, or graphically (see the Plot Edit Toolbar, Plot Browser, and Property Editor in the figure's View menu). Here is just a taste of what you can do:

```
>> h = plot(t,sin(t))
>> set(h,'color','m','linewidth',2,'marker','s')
>> set(gca,'pos',[0 0 1 1],'visible','off')
```

Handles also make it easy to change entire batches of objects at once. For example, the following code makes all of the blue lines in the current figure have width equal to 3:

```
>> h = findobj(gcf,'type','line','color','b');
>> set(h,'linewidth',3)
```

Because of handles, plots in MATLAB can be created in a basic form and then modified to look exactly how you want. However, it can be useful to change the default property values that are initially used to render an object. You can do this by resetting the defaults at any level above the target object's type in the hierarchy of Figure 5.1. For instance, to make sure that all future text objects in the current figure have font size 10, enter

```
>> set(gcf,'defaulttextfontsize',10)
```

Table 5.3. *Common colors in RGB and as string abbreviations.*

Color	RGB vector	Abbreviation
Black	[0 0 0]	'k'
Red	[1 0 0]	'r'
Green	[0 1 0]	'g'
Blue	[0 0 1]	'b'
Yellow	[1 1 0]	'y'
Magenta	[1 0 1]	'm'
Cyan	[0 1 1]	'c'
White	[1 1 1]	'w'

All figures are also considered to be children of the virtual root object that has handle 0, so setting properties to this handle creates global defaults.

5.4 Color

The coloring of lines and text is easy to understand. Each object has a color property that can be assigned an RGB (red, green, blue) vector in which each of the three entries is between zero and one. The primary and secondary colors are shown in Table 5.3, along with their one-letter string abbreviations. Note that green does not show well against a white background; MATLAB often uses a darker green color given by [0 0.5 0].

Surfaces are colored differently. There are two aspects, both modifiable: coloring of the data points, and coloring of the patches or mesh lines between them. The data point colors are specified by a CData property of the surface object. This could be a specification of RGB values at each individual data point, which is called the **truecolor model**. It is best for photographs and bitmap images.

The more common model for abstract data, and the default when you specify no explicit color information, is the **indexed color model**. This model involves an interaction between the CData property of the surface, the CLim property of its parent axes, and the Colormap property of the axes' parent figure. The figure's Colormap is an $m \times 3$ array, with each row interpreted as RGB values. The CLim property is a vector [a b] that defines a linear transformation from the interval $[a,b]$ to the interval $[1,m]$. Thus, each value in the CData property is mapped according to the linear transformation, then rounded to the nearest integer in $\{1,2,\ldots,m\}$, which serves as a row index into the colormap to determine a color.

As with all graphics object properties, those named above can be changed using set. However, there are more commonly used alternatives. By default, the CData surface property is equal to the array of z coordinate values (ZData) for the surface, but it can be set by giving a fourth argument to the plotting functions surf and mesh. The CLim is set by default to include exactly the full range of data values, and it can be changed after the fact by using caxis. The default figure Colormap is named jet and varies from blue to green to red. It can be

Table 5.4. *Shading models for surfaces.*

shading flat	Each face or mesh segment has constant color, determined by one boundary point.
shading faceted	Flat shading is used for faces and black for edges.
shading interp	Color in each face or segment is found by linear interpolation of corners.

changed by using the `colormap` command. Any changes to these properties, however made, have instant effects on the colors:

```
>> [X,Y,Z] = peaks;          % some built-in data
>> surf(X,Y,Z)
>> colorbar                  % show data->color mapping
>> caxis                     % current CLim value of axes
ans =
-6.5466     8.0752

>> caxis([-8 8]), colorbar   % zero-level symmetry
>> colormap pink             % change figure colormap
>> colormap gray             % change figure colormap
>> colormap(flipud(gray))    % reverse ordering
```

You can specify explicit `CData` for a surface in the indexed model. One natural use of this is for functions of a complex variable:

```
>> [T,R] = meshgrid(2*pi*(0:0.02:1),0:0.05:1);
>> [X,Y] = pol2cart(T,R);
>> Z = X + 1i*Y;
>> W = Z.^2;
>> surf(X,Y,abs(W),angle(W)/pi)  % arg(W) for coloring
>> axis equal, colorbar
>> colormap hsv                   % periodic colormap
```

Between grid points, color is determined by **shading**. You can change the shading of a surface by calling `shading` after the surface is created. The available types are shown in Table 5.4. While interpolated shading makes much smoother colors and prettier pictures, it can be very slow to render, particularly on printers. In fact, it's often faster to interpolate the data yourself to a finer grid and print with flat shading. See `interp2` to get started on this.

5.5 Saving and exporting figures

It often happens that a figure needs to be changed long after its creation. You can simply save the figure-creation commands in a script (section 3.1), but this approach has drawbacks. If the plot data takes a long time to generate, rerunning the script will waste time. Also, any edits made through menus and buttons will be lost.

Instead, you can save a figure in a format that allows it to be recreated in future sessions. Just enter

```
>> saveas(gcf,'myfigure.fig')
```

to save the current figure in a file named myfigure.fig. (This is equivalent to Save as... on the figure's File menu.) Later you can enter open myfigure.fig to recreate it. Since graphics objects store the data that defines them (in the XData, YData, and ZData properties), this tactic is also a way of storing the data used to create the graph. You can even leave notes about the data in an object's UserData field.

Ultimately, you will want to use some figures in a presentation or publication, a process that involves **exporting** the figure to a graphics file.[9] In this process, what you see on the screen is not necessarily what you get in the file, because MATLAB tweaks graphs differently depending on the output device. The relevant changes are laid out as a GUI if you select Export setup... from a figure's File menu. This dialog also lets you save a collection of preferences as a named style, so the GUI might be the most convenient approach if it meets your needs. But all of the options it offers affect properties that can be set using the handles of the figure and its descendant objects, should you prefer to work with commands.

Here are some significant issues to consider when exporting a figure:

Size. At first the size of the output graphic might seem irrelevant, since any document preparation or presentation program will let you change the image's size in the final product. However, resizing after exporting scales everything in the figure, including text, sometimes leading to graphs with thin lines or comically small font sizes. You are better off using MATLAB to set figure size, line widths, and font sizes independently before exporting.

Format. The most important difference in graphics file formats is between **vector** formats (such as EPS, PDF, and EMF) and **bitmap** formats (such as GIF, JPEG, PNG, and TIFF). Bitmaps, which store a pixel-by-pixel snapshot, are ideal for photographs but less so for most other scientific applications. These formats fix the resolution of your image forever, whereas the resolution of your screen, your printer, and a publication's printer are all very different. Vector formats, which represent lines, text, and surfaces more abstractly, are usually a much better choice for data-driven graphics.

EPS (encapsulated PostScript) files are the best choice for inclusion by LaTeX documents. They contain bounding box information that allows LaTeX to know exactly how large the graphic is. If you wish to export from the command line, you can use

```
>> print -deps myfig
```

---

[9]Depending on what tools you use, it may also be possible to copy and paste images, which is often adequate, at least on some platforms. However, you still might need to set up the figure first in the same ways discussed here.

or, for color output,

```
>> print -depsc myfig
```

to create `myfig.eps` from the current figure. However, this method uses the `PaperPosition` property of the figure, not the position on the screen, to determine the size of the output graphic. The EPS format also works with Microsoft Word™ if you print the output on a PostScript printer. In this case it's handy to use

```
>> print -deps -tiff myfig
```

in which case Word will be able to show a crude preview version of the graph in the document onscreen.

An additional twist in LaTeX is that if the file is processed directly to PDF format using pdflatex, EPS graphics cannot be used. Instead, you should use PDF for vector graphics. Currently, MATLAB-generated PDFs do not seem to embed graphic bounding box information in a way that the graphicx package for LaTeX understands. To work around this limitation, you can use the widely available pdfcrop script, or save the file as EPS and use the free script epstopdf to convert it to PDF.

Color. Colored lines are automatically converted to black when saved in a non-color format, so you should distinguish them by other features, such as symbol or line style. The colors of surfaces may be converted to grayscale, which presents a problem. The default colormap has colors ranging from deep blue to deep red, and the hues signal real information about the surface. When these colors are converted to grayscale, the distinction between red and blue is not clear, and the shades become meaningless. For a grayscale surface plot, you should consider using `colormap(gray)` or `colormap(flipud(gray))`, whichever gives less total black, before exporting the figure, so that saturation corresponds to data. Finally, the edges of wireframe surfaces created by `mesh` are also converted to gray, often with poor results. Make all the lines black by entering `colormap([0 0 0])`.

5.6 Other common graphics techniques

To create an array of plots in one figure window, use `subplot`. For example:

```
for n = 1:6
  subplot(2,3,n), ezpolar(@(t) sin(n*t), [0 2*pi])
  xlabel(''), title(['r = sin(',int2str(n),'t)'])
end
delete(findobj(gcf,'type','text'))   % generic labels
```

The call `subplot(m,n,k)` selects the *k*th axes in an $m \times n$ array, counting from left to right, top to bottom.

Here is a way to make a dynamic graph or animation:

```
figure('doublebuffer','on')    % flash-free redraws
t = linspace(0,8*pi,800)';
for s = 0:0.01:1
  x = exp(-s*t).*cos(6*s*t+t);
  y = exp(-s*t).*sin(6*s*t+t);
  plot(x,y), axis([-1 1 -1 1])
  pause(0.01)
end
```

Graphics commands accumulate in a buffer and do not take effect immediately. This fact allows you to make multiple changes (for instance, data and title) between animation frames. You must use a `pause` or `drawnow` command whenever you want to force all pending graphics commands to be processed.

The animation technique is best when little time is needed to render each frame. An alternative is to make a movie of bitmap renderings of the frames, using `getframe` to store each frame. The movie can be played back with `movie` or converted to a video file using `movie2avi`.

Generally, when a graphical function encounters a NaN value in a list of points, it quietly omits drawing the point. This can be useful when combined with the technique of index masking (see section 6.3). For example, these commands plot a surface that is mathematically defined only over an L-shaped region:

```
>> [X,Y] = meshgrid( -1:1/18:1 );
>> L = membrane(1,18,8,8);
>> surf(X,Y,L)
```

For this graph, the z values outside the domain are set to zero, which is not bad since the function happens to be zero at the real boundary. But the graph is improved if we erase that part of the surface using NaN:

```
>> outside = (X < 0) & (Y > 0);    % logical index mask
>> L(outside) = NaN;               % note "scalar expansion"
>> surf(X,Y,L)
```

Exercises

5.1. On a single graph, make a plot of the functions sinh, cosh, and tanh for $-1 \le x \le 1$. Give each curve a different color and label them with text and arrows.

5.2. Recall the identity

$$e = \lim_{n \to \infty} r_n, \qquad r_n = \left(1 + \frac{1}{n}\right)^n.$$

Make a standard and a log-log plot of $e - r_n$ for $n = 5, 10, 15, \ldots, 500$. What does the log-log plot reveal about the asymptotic behavior of $e - r_n$ as $n \to \infty$?

5.3. Here are two different ways of plotting a sawtooth wave. Explain concisely why they behave differently:

```
>> x = [0:7;1:8];   y = [zeros(1,8);ones(1,8)];
>> subplot(121), plot(x,y,'b'), axis equal
>> subplot(122), plot(x(:),y(:),'b'), axis equal
```

(The first version is more mathematically proper, but the second is more likely to appear in print.)

5.4. Play the "chaos game." Let P_1, P_2, and P_3 be the vertices of an equilateral triangle. Start with a point anywhere inside the triangle. At random, pick one of the three vertices and move halfway toward it. Repeat indefinitely. If you plot all the points obtained, a very clear pattern will emerge. (Hint: This is particularly easy to do if you use complex numbers. If z is complex, then plot(z) is equivalent to plot(real(z),imag(z)).)

5.5.   a. Generate 100 random matrices using randn(100), and plot all of their eigenvalues as dots in the complex plane on one graph. (Thus, you should see 10,000 dots.) Use axis equal to make the aspect ratio one-to-one. You should see a fairly striking result.

    b. Repeat the experiment with 100 random complex matrices of the form complex(randn(100),randn(100)). You should be able to see one very clear qualitative difference between the previous case and this one.

5.6. Make surface plots of the following functions over the given ranges:

1. $(x^2 + 3y^2)e^{-x^2 - y^2}$, $-3 \le x \le 3$, $-3 \le y \le 3$.
2. $-3y/(x^2 + y^2 + 1)$, $|x| \le 2$, $|y| \le 4$.
3. $|x| + |y|$, $|x| \le 1$, $|y| \le 1$.

5.7. Make contour plots of the functions in the previous exercise.

5.8. Make a contour plot of

$$f(x, y) = e^{-(4x^2 + 2y^2)}\cos(8x) + e^{-3((2x+1/2)^2 + 2y^2)}$$

for $-1.5 < x < 1.5$, $-2.5 < y < 2.5$, showing only the contour at the level $f(x, y) = 0.001$. You should see a friendly message.

5.9. Plot the surface represented by

$$x = u(3 + \cos(v))\cos(2u), \ y = u(3 + \cos(v))\sin(2u), \ z = u\sin(v) - 3u$$

for $0 \le u \le 2\pi$, $0 \le v \le 2\pi$.

Chapter 6

Advanced Techniques

MATLAB software provides a high-level language that allows one to think fairly abstractly about matrices, vectors, and functions. Sometimes, though, we must pay more attention to concrete concerns such as execution speed and programming complexity. In these contexts, certain patterns or habits prove to be more useful than others.

Sooner or later you will hear that MATLAB is "too slow." MATLAB does tend to prioritize development time over execution time. When your code is too slow to please you, start by profiling it (see section 3.5) to find out where the bottlenecks are. With more careful coding of time-intensive spots, you might recover huge gains in efficiency.[10] This chapter suggests some of the most common ways to find substantial savings.

Similarly, while vectors and matrices are a good paradigm for many situations, there are other tasks for which they can be found wanting. In this chapter we see other, more convenient, ways to manipulate data.

6.1   Memory preallocation

MATLAB hides the tedious process of allocating memory for variables. This generosity, while often highly convenient, can cause you to waste a lot of runtime. Consider the following code, which implements Euler's method for the vector differential equation $y' = Ay$ and stores the value at every time step:

---

[10]In a pinch, you can write a time-consuming subroutine in C or Fortran and link the compiled code into MATLAB. See the online help under "external interfaces."

```
tic
A = rand(200);
y = ones(200,1);
dt = 0.001;
for n = 1:(1/dt)
  y(:,n+1) = y(:,n) + dt*A*y(:,n);
end
toc
```

This takes about 0.93 seconds on a 2007-built desktop PC. Almost all of this time, though, is spent on a noncomputational task.

When MATLAB encounters the statement `y = ones(200,1)`, representing the initial condition, it asks the operating system for a block of memory to hold 200 floating-point numbers. Upon the first execution of the loop, it becomes clear that we actually need space to hold 400 numbers, so a new block of this size is requested. On the next iteration, this also becomes obsolete, and more memory is allocated. The little program above requires 1001 individual memory allocations of increasing size!

Changing the second line to `y = ones(200,1001);` changes none of the mathematics but does all the required memory allocation at once. This step is called **preallocation**. With preallocation the program takes about 0.075 seconds on the same computer as above, increasing the speed by a factor of 12.

6.2   Vectorization

Vectorization refers to the avoidance of `for` and `while` loops. As an example, suppose x is a column vector and you want to compute a matrix D such that $d_{ij} = x_i - x_j$. The standard implementation would involve two nested loops:

```
n = length(x);
D = zeros(n);    % preallocation
for j = 1:n
  for i = 1:n
    D(i,j) = x(i) - x(j);
  end
end
```

The loops could be written in either order here. But the innermost loop is easily replaced by a vector operation:

```
n = length(x);
D = zeros(n);    % preallocation
for j = 1:n
  D(:,j) = x - x(j);
end
```

We can get rid of the remaining loop, too, by upgrading from vectors to two-dimensional arrays. This is a bit more subtle:

```
n = length(x);
X = x(:,ones(n,1));      % copy columns to make n by n
D = X - X.';
```

The second line here is a trick that was introduced in section 2.2 (and in MATLAB circles is called "Tony's trick"). Note too the use of . ' in the last line for compatibility with a complex input.

At this point you might ask, why vectorize? There are two answers: speed and style. Neither is a simple matter. Until around 2002, careful vectorization almost always yielded tremendous improvements in execution speed. But the situation has changed somewhat due to a technique called **JIT acceleration**. Acceleration, which is applied automatically, can remove the speed penalty that MATLAB traditionally experienced with loops. While not every loop can be optimized, it's clear that code vectorization is not critical for speed in every case. For example, with $n = 1000$ the times in milliseconds for each of the three methods above were 0.025, 0.010, and 0.025, respectively. In this case a medium level of vectorization proved to be the fastest.

Style is a subjective matter, of course. To programming veterans who are new to MATLAB, multiply nested loops over scalar operations seem natural and perhaps inevitable. However, MATLAB makes it easy to act on vectors and matrices, and not on their individual entries, as atomic objects. The superiority of A*B to a triply nested loop to compute a matrix product is obvious; many problems seem to land between extremes. In the codes above, for instance, each version is shorter than the one before it. Yet while the second is no less clear than the first, the third version requires some thought when first encountered. Unlike the first two versions, the third is also not easily adjusted to account for the obvious antisymmetry in the result.

Another classic illustration of the trade-offs involved with vectorization comes from Gaussian elimination. Here is a basic textbook implementation without any vectorization:

```
n = length(A);
for k = 1:n-1
  for i = k+1:n
    s = A(i,k)/A(k,k);
    for j = k:n
      A(i,j) = A(i,j) - s*A(k,j);
    end
  end
end
```

Again we can start vectorization with the innermost loop, on j. Each iteration of this loop is independent of all the others. This parallelism is a big hint that we can use a vector operation instead:

Table 6.1. *CPU times, in milliseconds per factorization, for $n \times n$ Gaussian eliminations using three different levels of vectorization.*

	$n = 100$	200	300	400	500
Three loops	23	143	369	743	1364
Two loops	50	171	382	837	1584
One loop	6.2	43.8	258	962	2188

```
n = length(A);
for k = 1:n-1
   for i = k+1:n
      s = A(i,k)/A(k,k);
      cols = k:n;
      A(i,cols) = A(i,cols) - s*A(k,cols);
   end
end
```

The new version makes the algorithmic idea of a row operation much more apparent, and, in my view, is clearly preferable. However, the innermost of the remaining loops is also vectorizable:

```
n = length(A);
for k = 1:n-1
   rows = k+1:n;
   cols = k:n;
   s = A(rows,k)/A(k,k);
   A(rows,cols) = A(rows,cols) - s*A(k,cols);
end
```

You have to flex your linear algebra muscles a bit to see that the vector outer product in the next-to-last line is appropriate. This is an interesting insight, but does not lead to an unquestionable improvement in style.

To compare the speed of these three versions, each was run 20 times for different values of the matrix size n on a 2007 PC using MATLAB 7.7. The results, per factorization in milliseconds, are given in Table 6.1. The results of this experiment advise against making a uniform recommendation about vectorization based on execution speed! The experiment also points out the pitfalls of classical thinking about the performance of algorithms solely in terms of floating-point operation counts: none of the above rows is well represented as an $O(n^3)$ function.

An uncontroversial application of vectorization is the use of array operators and utility functions in the evaluation of a mathematical expression. In the following comparisons, it's hard to defend the looped versions as being superior to their vectorized counterparts, once you understand the functions being used:

```
y = t.*sin(t.^2);
```
```
y = zeros(size(t));
for i = 1:length(t)
    y(i) = t(i)*sin(t(i)^2);
end
```

```
dz = diff( z([1:end 1]) );
```
```
dz = zeros(size(z));
for j = 1:length(z)-1
    dz(j) = z(j+1)-z(j);
end
dz(end) = z(end)-z(1);
```

```
e = 1+sum(1./cumprod(1:20));
```
```
e = 1; p = 1;
for j = 1:20,
    p = p*j;
    e = e + 1/p;
end
```

The bottom line is that loops should not be written carelessly. Writing code that works at a vector level is easy and natural in most cases. But if profiling indicates that a lot of time is being spent in a place where the level of vectorization is selectable, experimentation may be the only way to see what level is best. For leads on more sophisticated types of vectorization, look for accumarray, arrayfun, and bsxfun in the online help.

6.3 Masking

A special type of vectorization is called **masking**. Let's say that we have a vector x of values at which we want to evaluate the piecewise-defined function

$$f(x) = \begin{cases} 1+\cos(2\pi x), & |x| \leq \frac{1}{2}, \\ 0, & |x| > \frac{1}{2}. \end{cases}$$

Here is the standard loop method:

```
f = zeros(size(x));
for j = 1:length(x)
    if abs(x(j)) <= 0.5
        f(j) = 1 + cos(2*pi*x(j));
    end
end
```

The shorter way is to use a mask:

```
f = zeros(size(x));
mask = (abs(x) < 0.5);
f(mask) = 1 + cos(2*pi*x(mask));
```

The mask is a logical index into x (see page 17). You could refer, if needed, to the unmasked points by using ~mask.

Consider a new version of the sum-of-primes idea from page 32. Here's how we could count the number of primes less than 100 and add them up:

```
isprm = isprime(1:100);    % which are prime?
sum( isprm )               % how many primes
sum( find(isprm) )         % sum the primes
```

Here find converts a logical index into an absolute one, i.e., gives a vector of the prime numbers.

6.4 Scoping exceptions

Once in a while, the scoping rules for functions get in your way. Although you can virtually always do what you need within the rules, it's nice to know how to bend them.

The least useful and most potentially troublesome violation of scoping comes from **global variables**. Any variable may be declared global before it is assigned a value for the first time in its current scope. After this, any other workspace may also declare the variable to be global and access or change its value. The trouble with global variables is that they do not scale well to large or even moderately sized projects; name clashes and unexpected behaviors tend to creep in quickly. At one time global values were occasionally more or less necessary, so you may still find examples of code that uses them. However, all the legitimate uses are now achieved by more desirable and stable mechanisms. In particular, global variables should not be used to pass parameters between functions. Instead, use the techniques described in sections 4.1 and 4.2.

A more interesting exception to the scoping rules is the **persistent** variable. When a function declares a variable to be persistent, the variable's value is preserved between invocations to that particular function. While this mechanism does require a little care to ensure correct use, it is far less general than a global variable, because the persistent variable is still visible only within the workspace of the function that declares it.

Persistent variables can be used to record information about a function's internal state, or to preserve costly preliminary results that can be reused later. Consider this example for computing the Fibonacci numbers:

```
function y = fib(n)

persistent f
if length(f) < 2, f = [1 1]; end
for k = length(f)+1:n
  f(k) = f(k-2) + f(k-1);
end
y = f(1:n);
```

The first time this function is called, f will be empty.[11] Thus, f will then be assigned the first two values in the sequence. After this, f is extended as needed to get the first n values, and the sequence is returned to the caller. In future calls to fib, any previously computed members of the sequence are simply accessed rather than recomputed. If the function were called requesting sequence lengths of $n_1, n_2, \ldots, n_k$, the work would be proportional to $\max n_i$ rather than $\sum n_i$. The same effect could, of course, be achieved by having f as an input argument, but that makes the calling function responsible for handling data that may be relevant only within fib. The approach here is more self-contained.

6.5 Strings

A MATLAB string is enclosed in single forward quotes. If you want a quote character within a string, use two consecutive quote characters, as in 'It''s Cleve''s fault'.

A string is really just a specially flagged row vector of character codes, and thus strings can be concatenated using matrix syntax:

```
>> str = 'Hello world';
>> str(1:5)
ans =
Hello

>> double(str)
ans =
    72   101   108   108   111   ...   108   100

>> char(ans)
ans =
Hello world

>> ['Hello',' ','world']
ans =
Hello world

>> ['Hello'; 'world']
ans =
Hello
world
```

Vertical concatenation is not so straightforward in general, because each row of an array has to have the same length. You can use char instead to automatically pad strings with blanks before concatenation. However, if your intent is really to create a collection of strings, a better mechanism is a cell array (section 6.6).

---

[11]Unlike other variables, persistent variables are given an initial value—specifically, the empty matrix.

There are many string handling functions; see the help on `strfun`. Here are a few:

```
>> upper(str)
ans = HELLO WORLD

>> strcmp(str,'Hello world')
ans =
     1

>> findstr('world',str)
ans =
     7
```

Of particular interest is converting between numbers and their character representations. You can convert a string such as `'3.14'` into its numerical meaning by using `str2num` or `str2double`. Note the important difference between `double`, which converts to character codes, and `str2double`, which interprets the string as a number!

For the inverse operation of converting a number to a string, there are `num2str` and `sprintf`. Both accept C-style format strings with data conversion specifiers (but unfortunately, without extension to complex numbers). In `num2str`, a single conversion specifier is applied to all the elements of an array, while the more flexible `sprintf` can apply multiple conversion specifiers cyclically through all the data given, proceeding in the usual row-first ordering through arrays. For output to the screen or a file rather than a string, use `fprintf`.

A common use of `sprintf` is to create strings, such as file names, that have consecutive integers in them, as in this example:

```
for n = 1:12
  fn = sprintf('mydata_%.2i',n);
  load(fn), data(:,:,n) = A;
end
```

This will load each of the data files named mydata_01.mat, …, mydata_12.mat, storing a matrix called A as layers of a three-dimensional array. Note that

```
        load mydata_01     and     load('mydata_01')
```

are functionally identical, but the second is more suitable when the file name is stored in a variable. The same "command/function duality" holds for every MATLAB command that takes extra arguments separated by spaces.

Another common need is to create formatted tables of output. For example, this script prints out successive Taylor approximations for $e^{1/4}$:

```
x=0.25;   n=1:6;   c=1./cumprod([1 n]);
for n=1:7,   T(n)=polyval(c(n:-1:1),x);   end
fprintf('\n       T_n(x)        |T_n(x)-exp(x)|\n');
fprintf('-------------------------------------\n');
fprintf('%15.12f       %8.3e\n', [T;abs(T-exp(x))]  )
```

The result:

```
      T_n(x)           |T_n(x)-exp(x)|
  ---------------------------------
  1.000000000000       2.840e-01
  1.250000000000       3.403e-02
  1.281250000000       2.775e-03
  1.283854166667       1.713e-04
  1.284016927083       8.490e-06
  1.284025065104       3.516e-07
  1.284025404188       1.250e-08
```

6.6 Cell arrays

Collecting objects of different sizes is a common chore. For instance, suppose you want to tabulate the Chebyshev polynomials $T_0 = 1$, $T_1 = x$, $T_2 = 2x^2 - 1$, $T_3 = 4x^3 - 3x$, and so on. In MATLAB one expresses a polynomial as a vector of its coefficients, so the length of the vector depends on the degree of the polynomial. For a collection of consecutive Chebyshev polynomials starting from T_0, we could use a triangular array, but this is neither convenient nor general.

Cell arrays are a mechanism for gathering dissimilar objects into one variable. They are indexed like regular numeric arrays, but their elements can be anything, including other cell arrays. A cell array is created or referenced using curly braces { } rather than parentheses. Cell arrays can have any size and dimension, and their elements do not need to be of the same size or type. Because of their generality, cell arrays are mostly just containers; they do not support any sort of arithmetic.

```
>> str = { 'Goodbye', 'cruel', 'world' }
str =
    'Goodbye'     'cruel'     'world'

>> str{2}
ans =
cruel

>> T = cell(1,9);
>> T(1:2) = { [1], [1 0] };
>> for n=2:8,   T{n+1}=[2*T{n} 0] - [0 0 T{n-1}];   end
```

```
>> T
T =
  Columns 1 through 5
    [1]      [1x2 double]    ...     [1x5 double]

  Columns 6 through 9
    [1x6 double]      [1x7 double]    ...      [1x9 double]

>> T{4}
ans =
      4       0      -3       0
```

The second example above points out a subtlety in extracting elements from cell arrays. The assignment reference T(1:2) refers to a subarray extracted from the cell array T; thus, the right-hand side member of the assignment is a 1×2 cell array to match. The reference T{n+1}, however, refers to a single element of a cell array—in this case, a vector. To summarize,

```
>> T{4}
ans =
      4       0      -3       0

>> T(4)
ans =
    [1x4 double]
```

Thus, the right-hand side member of the loop assignment must not be enclosed in braces. Doing so would not cause an error in the first loop pass—the element T{3} would itself be a cell array with one element—but it would cause an error in the second pass, when that cell array is illegally concatenated with a vector.

What, then, should we expect from a reference to T{4:5}, which should refer to two vectors? The answer is that MATLAB acts as though those two vectors had just been entered in the command line at that spot:

```
>> T{4:5}
ans =
      4       0      -3       0

ans =
      8       0      -8       0       1
```

This syntax proves to be surprisingly useful, in particular as the idiom T{:}, which is interpreted as a comma-separated list of the elements of T. For example, using the str cell array created above, we get

```
>> char(str{:})    % = char('Goodbye','cruel','world')
ans =
Goodbye
cruel
world
```

When combined with `cat` (the command form of array concatenation), this syntax can be used to convert cells containing compatibly sized data into numeric array form:

```
>> c = { [3 4], [5 6] };
>> cat(1, [1 2], c{:} )
ans =
     1     2
     3     4
     5     6

>> cat(2, [1 2], c{:} )
ans =
     1     2     3     4     5     6

>> e = {};  cat(2, [1 2], e{:} )
ans =
     1     2
```

Observe that for the empty cell array e, the expression e{:} was not only empty, but it also gobbled up the comma in front of it that would otherwise have caused a syntax error. The inverse operation of converting an array into cell form is done by num2cell:

```
>> num2cell(1:6)
ans =
    [1]     [2]     [3]     [4]     [5]     [6]

>> num2cell(1:6,2)
ans =
    [1x6 double]
```

The second argument of `num2cell` specifies which dimension to "pack" into a cell; it defaults to 1.

The special cell array `varargin` is used to pass optional arguments into functions. In this way, you can write functions that have sensible default behaviors which may be overridden. For example, suppose we want to write a function that returns the color specification for blue, in either the RGB color model (by default) or the hue-saturation-value (HSV) model:

```
function b = blue(varargin)

if nargin < 1
  varargin = {'rgb'};
end

switch(varargin{1})
  case 'rgb'
    b = [0 0 1];
  case 'hsv'
    b = [2/3 1 1];
  otherwise
    error('Unrecognized color model.')
end
```

Both `blue` and `blue('rgb')` return the first option, while `blue('hsv')` returns the second. Another way to use `varargin` is when an arbitrary number of inputs might be appropriate. For instance, consider:

```
function s = add(s,varargin)

for n = 1:nargin-1
  s = s + varargin{n};
end
```

Here, the first input argument is assigned to the name `s`, and all others are assigned to the cell `varargin`. The keyword `nargin` always returns the total number of inputs, both named and optional. To provide similar functionality for output arguments, you can use `varargout` and `nargout`.

6.7   Structures

Structures are essentially cell arrays that are indexed by name rather than by number.

Say you are keeping track of the grades of students in a class. You might start by creating a student structure as follows:

```
>> student.name = 'Moe';
>> student.homework = [10 10 7 9 10];
>> student.exam = [88 94];
>> student
student =
         name: 'Moe'
     homework: [10 10 7 9 10]
         exam: [88 94]
```

The name of the structure is `student`. Data is stored in the structure according to named **fields**, which are accessed using the dot notation. The field values can be anything.

Let's add another student:

```
>> student(2).name = 'Curly';
>> student(2).homework = [4 6 7 3 0];
>> student(2).exam = [53 66];
>> student
student =
1x2 struct array with fields:
    name
    homework
    exam
```

Now you have an array of structures. This array can have any size and dimension. All elements of the array must have the same fields, but the values occupied by those fields do not have to be compatible in any way across different elements of the structure array. For instance,

```
>> student(2).homework = 'missing'
```

would be a syntactically valid continuation of the above.

In parallel with the cell array syntax `c{:}`, you can use the array and field names alone to create comma-separated lists of all the entries in the array. For example:

```
>> roster = {student.name}
roster =
    'Moe'       'Curly'

>> hw = cat(1, student.homework )
hw =
    10    10     7     9    10
     4     6     7     3     0
```

In addition, you can refer to a field whose name is stored as a string variable using a modified syntax:

```
>> score = 'exam';
>> cat(1, student.(score) )
ans =
    88    94
    53    66
```

Exercises

6.1. Rewrite `trap` or `simp` (page 43) so that it does not use any loops. (Hint: Set up a vector of all x values and evaluate `f` for it. Then set up a vector of weighting coefficients and use an inner product.)

6.2. Consider again the pairwise-differencing code variants beginning on page 60. Explore the issue of whether the order of the loops in the first variant, or the use of row rather than column vectors in the second variant, affects the execution time.

6.3. Suppose x is a column vector. Compute, without using loops or conditionals, the matrix A given by

$$a_{ij} = \begin{cases} \dfrac{1}{(x_i - x_j)^2} & \text{if } i \neq j, \\ 1 & \text{otherwise.} \end{cases}$$

(One way to do this uses direct assignment to the diagonal elements of A. Using row/column style indices, this is rather tricky, but see Exercise 2.6.)

6.4. The $n \times n$ matrix

$$A_n = \begin{bmatrix} 1 & 0 & 0 & \cdots & 1 \\ -1 & 1 & 0 & \cdots & 1 \\ -1 & -1 & 1 & \cdots & 1 \\ \vdots & & \ddots & \ddots & \vdots \\ -1 & -1 & \cdots & -1 & 1 \end{bmatrix}$$

is well known for causing (a very rare!) numerical instability in Gaussian elimination for solving systems of equations. Using some combination of commands such as `ones`, `tril`, `eye`, indexed assignment, and array concatenation, create the matrix A_{50} without resorting to any loops.

6.5. Reconsider the function `chebeval` (page 34) that evaluates Chebyshev polynomials at multiple points. Write the function so that it performs as efficiently as you can make it. Among other things, you have to choose whether to use the definition or the recursion.

6.6. Consider "shuffling" a vector of integers from 1 to 52 using a physical interpretation of a card shuffle. Divide the cards into two equal stacks, and merge the stacks together such that each time a pair of cards is to fall off the "bottom" of each stack, a random decision is made as to which falls first. A loop-based implementation would be

```
function s = shuffle(x)
n = length(x)/2;
s = [];
for i = 1:n
  if rand(1) > 0.5
    s = [s x([2*i-1 2*i])];
  else
    s = [s x([2*i 2*i-1])];
  end
end
```

Rewrite this function without loops. (This can be done in as few as four statements using resizing tricks.) It can be interesting to start with a perfectly ordered deck and see how many shuffles it takes to "randomize" it. One crude measure of randomness is the (1,2) element of corrcoef(1:52,v), which is expected to be zero if v is random.

6.7. Rewrite the function entropy on page 35 without any loops using sort, diff, find, and (perhaps) sum.

6.8. Different Fibonacci sequences can be produced by changing the first two members of the sequence. Rewrite fib from page 64 so that it accepts these seed values and recomputes the sequence *only* when necessary.

6.9. Write a function capital that accepts arbitrarily many string inputs and capitalizes each one using the built-in function upper. You can return either a cell array of the new strings, or a list of the individual strings using varargout.

6.10. Write a function

```
function f = pcfun(int_1,val_1,...,int_N,val_N)
```

that returns a callable function f such that $f(x)=$val_k if x is in the interval int_k, specified as a vector $[a\ b]$, and not in any previous int_j for $j < k$, and such that $f(x) = 0$ if x is not in any of the intervals. The f you return should be a handle to a nested function within pcfun. To increase the challenge, allow a different default value to be specified at the end of the list.

6.11. Using the examples in section 6.7 as a model, write a function

```
function [mu,sigma] = gradestat(student,asgnt)
```

that computes grade averages and standard deviations for a field named in the input argument asgnt.

Chapter 7

Scientific Computing

Certain computational problems appear repeatedly in applications motivated by science and engineering. Naturally, the MATLAB software is thoroughly equipped with ways to handle such problems. This chapter presents quick surveys of some of these methods. The goal of the chapter is not to make you a numerical master chef or even to give you "recipes" for numerics. Rather, it's to explain how to order from the menu. A solid mathematical understanding of the underlying methods would require at least a year of undergraduate numerical analysis. However, even complex problems can often be broken down into linked elements for which MATLAB provides good functions for solving.

7.1   Linear algebra

The most common tasks in numerical linear algebra are solving linear systems and finding matrix decompositions, with eigenvalues and singular values being the most prominent examples. Broadly speaking, the methods for these problems break down into **direct methods**, which are best for matrices up to size (perhaps) in the few thousands, and **iterative methods**, which are useful for sparse and truly large matrices. Iterative methods are discussed in section 7.2.

Linear systems are most often given in one of two forms: $Ax = b$, where A is a square matrix, or in the **overdetermined least-squares** problem min $\|b - Ax\|_2$, where A is a matrix with more rows than columns. Mathematically, the square case is solved (when possible) by using the inverse A^{-1}, and the rectangular case is solved by a generalization known as the **pseudoinverse**. However, in numerical computation an actual matrix inverse is computed only rarely in practice. When numerical analysts speak of "inverting" a matrix, what's meant is using appropriate numerical algorithms that accomplish an equivalent task.

In MATLAB, the method of choice in both square and rectangular cases is the backslash operator, used in the form x=A\b for both types of problems. For square matrices, in fact, the idiom A\ is mathematically equivalent and computationally superior to left-multiplication by A^{-1}. When A is not square, the same idiom represents left-multiplication by the pseudoinverse of A. Similarly, to multiply on the right by A^{-1} or the pseudoinverse, you can use a forward slash, as in /A. The notational idea in both cases is that A is in the "denominator," while respecting the fact that matrix multiplication is not commutative. The generality of the notation is aesthetically pleasing but dangerous when used carelessly. For example, when A is $n \times n$ and b is $n \times 1$, then both A\b and b\A are syntactically and mathematically defined but utterly different!

The backslash operator does not apply only a single algorithm. It represents an expert system that applies the best of several methods depending on detectable properties of A such as sparsity, triangularity, and positive definiteness. These are spelled out in the documentation page for mldivide. One consequence of the complexity of this operator is that you may need to set up A properly to get the fastest solution: for instance, if A is tridiagonal, you should create it in or convert it to sparse format to get an $O(n)$ solution time. See section 2.5 for more on sparse matrices.

If you need to solve multiple systems $Ax_i = b_i$, for the known vectors $b_1, \ldots, b_k$, you can use the identity

$$\begin{bmatrix} b_1 & b_2 & \cdots & b_k \end{bmatrix} = \begin{bmatrix} Ax_1 & Ax_2 & \cdots & Ax_k \end{bmatrix} = A \begin{bmatrix} x_1 & x_2 & \cdots & x_k \end{bmatrix}.$$

The following code demonstrates a case with $k = 3$:

```
>> A = magic(5);
>> B = ones(5,3);   B(2,2)=0;   B(4,3)=0;
>> X = A\B
X =
      0.0154    -0.0358     0.0142
      0.0154     0.0527     0.0027
      0.0154     0.0123     0.0123
      0.0154     0.0219    -0.0281
      0.0154     0.0104     0.0604

>> A*X
ans =
      1.0000     1.0000     1.0000
      1.0000    -0.0000     1.0000
      1.0000     1.0000     1.0000
      1.0000     1.0000     0.0000
      1.0000     1.0000     1.0000
```

Often, however, the right-hand side vectors b_i are not all known at once, because they are produced as part of an iteration. In such a situation, the backslash idiom is much less useful, because it has to repeat costly work on each iteration. Instead, you need to first compute a separate factorization. For instance, the well-known inverse iteration for an eigenvector of A might look like this:

```
A = randn(400)/20;
[L,U] = lu(A);
```

```
v = randn(400,1);   v = v/v(1);
for k = 1:50
   v = U \ (L\v);
   v = v/v(1);
end
```

Observe that U\(L\v) is equivalent to $U^{-1}(L^{-1}v) = (LU)^{-1}v = A^{-1}v$. Because L and U have a special structure (triangular plus permutation), applying the backslash to them requires only triangular substitutions, which are much faster than the factorization step. Matters become even more complicated when A has one of the special properties that can be detected by the backslash. At this point one really needs to know some numerical linear algebra to exploit the structure of the problem.

Two of the most commonly desired matrix factorizations are the eigenvalue and singular value decompositions (SVDs). These are found using eig and svd, respectively. Both functions have multiple output formats, as demonstrated below:

```
>> A = magic(5);
>> lambda = eig(A)
lambda =
    65.0000
   -21.2768
   -13.1263
    21.2768
    13.1263

>> [V,D] = eig(A);
>> v1 = V(:,1)
v1 =
   -0.4472
   -0.4472
   -0.4472
   -0.4472
   -0.4472

>> norm(A*V-V*D)
ans =
   6.9390e-014

>> A = magic(5);
>> sigma = svd(A)
sigma =
    65.0000
    22.5471
    21.6874
    13.4036
    11.9008
```

```
>> [U,S,V] = svd(A);
>> norm(A-U*S*V')
ans =
   8.7934e-014
```

The differences are more than pure cosmetics, as the single-output forms can be substantially faster than the full versions. In addition, the syntax svd(A,0) produces a reduced-size variant of the SVD that is often more appropriate for rectangular matrices.

7.2   Iterative linear algebra

The methods of section 7.1 are best for matrices of size up to around a thousand, or perhaps tens of thousands for some sparse linear systems. For truly large problems, one turns to iterative methods, which produce sequences of improving estimates. Iterative methods do not factor or modify the given matrix, which is ideal for sparse problems, and they are able to exploit other problem structures, such as the existence of an approximate inverse (an idea known as **preconditioning**).

For linear systems, there are many choices of iterative methods. The focus in MATLAB is on the projective or **Krylov subspace** methods (see the help for sparfun), as opposed to older methods such as Gauss–Seidel or successive over-relaxation (SOR). If the system matrix is positive definite, the usual choice is the **conjugate gradient** method, implemented by pcg. For more general matrices, bicgstab and gmres tend to be quite popular, but there is no single best method.

Here is gmres at work:

```
>> A = eye(400) + 0.5*randn(400)/20;
>> b = randn(400,1);
>> x = gmres(A,b,10,1e-10);
gmres(10) converged at outer iteration 4 ...

>> norm(b-A*x)/norm(b)
ans =
   6.8167e-011
```

In practice the choice of method is often less critical than the choice of a precon-ditioner, which is a matrix M that is somehow close to A yet allows very fast solutions of $Mx = b$.

For matrix decompositions the choices are simpler: eigs for eigenvalues and sigs for singular values. Both are preferred to the power iteration and related classical methods often presented in textbooks. Usually one does not seek the full decomposition (which might be impractical to even store), but rather a few of the largest or smallest values and perhaps their associated vectors. Those capabilities are available, as explained in the online help.

All of these iterative methods have a surprising additional flexibility: they don't require matrices! They can instead be given a function that returns the vector

Au for any given vector *u*. This is more useful than it may first seem, for example, in applications where one can use multipole expansions, wavelets, or other fast approximate methods for the matrix-times-vector operation. This is a common mode of operation in many of the very largest problems.

7.3 Rootfinding

One of the oldest and most common numerical problems is trying to find a root or zero of a scalar function $f(x)$ of one variable. (This formulation is equivalent to solving $f(x) = c$ for any constant c, since in that case we can find a root of the function $f(x) - c$.) Usually one learns Newton's method for this problem, but it turns out not to be the best general-purpose method, owing to its unreliable convergence and its requirement of an explicit derivative calculation.

The MATLAB function `fzero` is much better than a handwritten algorithm for most rootfinding problems. It requires you to provide a function that computes $f(x)$ for any given x, and either a starting guess for the root or an interval on which f changes sign. Providing an interval, when this is feasible, can often greatly speed up the convergence. As an example, we try both approaches using a syntax that shows the progress of iterations. First, we try with a starting guess only:

```
>> f = @(x) 1 + 1./(1-x) + 4./(2-x);
>> opt = optimset('disp','iter');
>> fzero(f,1.5,opt)
Search for an interval around 1.5 ...
 Count        a         f(a)          b          f(b)
   1         1.5          7          1.5           7
   3        1.4576      6.1888      1.5424        7.898

    ...
  14        1.1606     -0.46184      1.74        15.033

Search for a zero in the interval [1.16059, 1.74]:
 Func-count     x          f(x)
   14        1.16059    -0.461842
   15        1.17786     0.242895

    ...
  21        1.17157   8.88178e-016

Zero found in the interval [1.16059, 1.74]

ans =
   1.1716e+000
```

(The output has been edited to save space.) Here two-thirds of the iterations were devoted just to finding an interval that contains a root. If we provide such an interval to start with, much of the work is saved, even if the interval is not small:

```
>> fzero(f,[1.01 1.99],opt)
  Func-count        x              f(x)
      2                    1.01       -94.9596
      3                 1.19802        0.937661
              . . .
     11                 1.17157     8.88178e-016

Zero found in the interval [1.01, 1.99]

ans =
   1.1716e+000
```

In this particular example, the execution time difference is negligible, but in many problems each evaluation of the function f might itself represent a time-consuming computation, such as integration of a differential equation.

Roots of polynomials are handled differently, using the function roots, which actually solves an equivalent matrix eigenvalue problem. It's far faster and more reliable than trying to find roots by Newton-style methods. Be aware, though, that polynomial roots can be very sensitive to perturbations in the coefficients.

For solving multidimensional systems of nonlinear equations of the form $F(x) = 0$, MATLAB offers fsolve in the Optimization Toolbox. You need only supply a function for computing F and a starting point, but you might greatly improve the convergence if you also supply a function computing the Jacobian matrix $J(x)$ of F, or at least giving the sparsity pattern of J. There are several algorithms to choose from, so read the documentation if you have trouble on a particular problem. All of the methods expect F to be reasonably smooth.

7.4   Optimization

To find the minimum of a scalar function $f(x)$ of one variable, use fminbnd. It requires a function computing f and an interval on which to optimize it:

```
>> format long
>> f = @(x) exp(3*x.^2-2*x);
>> fminbnd(f,0,2)
ans =
   0.333329150935613

>> fminbnd(f,0,2,optimset('tolx',1e-10))
ans =
   0.333333328332764
```

To maximize $f(x)$, minimize $-f(x)$.

For minimization of functions with multiple variables, there are two main choices: fminunc (in the Optimization Toolbox) and fminsearch. For smooth

functions, `fminunc` is usually faster, but `fminsearch` does not use any derivative information and may be better for nonsmooth problems. In the special case of
nonlinear least squares,

$$f(x) = f_1(x)^2 + f_2(x)^2 + \cdots + f_m(x)^2,$$

one should use `lsqnonlin` instead. (In fact, `lsqnonlin` is one of the methods
`fsolve` uses for solving nonlinear systems.)

   Optimization in engineering often comes with side constraints on the variables. None of the multidimensional methods named here respects such constraints. See the Optimization Toolbox for methods for several variants of such
problems.

7.5 Data fitting and interpolation

A common task is to make sense of data from noisy observations of a presumably
simple relationship. Here noise can refer to experimental error or uncertainty, or to
neglected higher-order effects. In data fitting one posits a form for the underlying
relationship involving some unknown parameters, then uses the data to find the
best parameters via an optimization.

   When the form of the relationship is a linear combination of basis functions
and the parameters are the coefficients, the resulting optimization is a linear least-
squares problem and can be solved using the backslash. For example,

```
>> t = (0:.01:3)';
>> y = 4+cos(t)-0.5*sin(t)+0.2*randn(size(t));   % data
>> A = [ones(size(t)) cos(t) sin(t)];   % basis functions
>> c = A\y
c =
    4.0033
    0.9890
   -0.4763
```

We can use an anonymous function (with linear algebra shortcuts) to turn the result
into a callable function of the independent variable:

```
>> f = @(t) [ ones(size(t)) cos(t) sin(t) ] * c;
>> max( abs(f(t)-y) )
ans =
    0.6093

>> ezplot(f,[0 3])
>> hold on, plot(t,y,'.')
```

The result is shown in Figure 7.1.

   When the functional form is a polynomial, the least-squares fitting process
is automated by `polyfit`. Certain other types of fit can be recast into a polynomial form. For example, the power law $y = ct^\alpha$ becomes the linear relationship

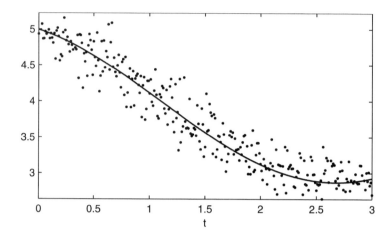

Figure 7.1. *Example of a least-squares fit to data.*

$\log y = \log c + \alpha \log t$ for the logs of the variables. A graphical data fitting tool for the most common linear fits is available from the Tools menu of every figure window.

   If the unknown coefficients of the fitting function appear nonlinearly, a non-linear optimization results. For instance, MATLAB ships with U.S. census data for the years 1790–1990. This data was collected every 10 years. To fit this data with an exponential curve of the form $c_1 + c_2 e^{c_3 t}$, you could proceed as follows (if you have access to the Optimization Toolbox):

```
>> load census
>> t = (cdate-1790)/10;
>> resid = @(c) sum((c(1)+c(2)*exp(c(3)*t)- pop).^2);
>> fminsearch(resid,[1 1 1])
ans =
  -39.1124    35.8209    0.1053
```

You can check graphically that this is a pretty good fit.

   A related problem is that of **interpolation**. The idea again is to transform discrete values into a function of a continuous variable, but here the function is constrained to actually pass through all the given data. This is most appropriate when one presumes that the data itself consists of samples of a function with some regularity properties.

   The primary function for interpolation in one variable is `interp1`. For example, here we sample a smooth function at eight random points, then compare a piecewise cubic interpolating polynomial to the original:

```
>> f = @(t) exp(sin(2*t));
>> x = [0;sort(rand(6,1));1];   y = f(x);
>> t = (0:0.2:1)';
>> [ interp1(x,y,t,'pchip'), f(t) ]
```

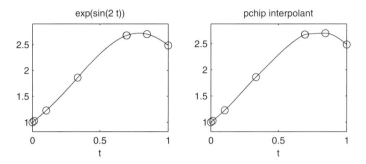

Figure 7.2. *One-dimensional interpolation.*

```
ans =
    1.0000    1.0000
    1.4873    1.4761
    2.0280    2.0490
    2.5194    2.5397
    2.6976    2.7171
    2.4826    2.4826
```

Calling `interp1` evaluates the interpolating polynomial at given points. If instead you want to create an interpolating function that can be evaluated, it is convenient to use a different form of the command with an anonymous function:

```
>> g = @(t) ppval( interp1(x,y,'pchip','pp'), t );
>> subplot(1,2,1), ezplot(f,[0 1]) % original
>> hold on, plot(x,y,'o')
>> subplot(1,2,2), ezplot(g,[0 1]) % interpolant
>> hold on, plot(x,y,'o')
```

The result is shown in Figure 7.2.

There are several different types of interpolants available with `interp1`: `'nearest'`, which produces a piecewise-constant staircase approximation; `'linear'`, the default, which is piecewise linear; `'spline'`, which is a piecewise cubic **spline** with not-a-knot conditions; and `'pchip'`, which is also piecewise cubic, but sacrifices some smoothness in order to preserve monotonicity and local maxima and minima in the result. You can also interpolate periodic data using `interpft`.

For interpolation of data depending on more than one variable, there are two families of commands: `interp2`, `interp3`, and `interpn` when the data values lie on grids (not necessarily equispaced), and `griddata`, `griddata3`, and `griddatan` when the data lies at scattered locations. In the latter cases, only `nearest` and `linear` types of interpolants are available.

7.6   Integration

Numerical approximation of definite integrals is called **quadrature** in one dimension and **cubature** if the integration domain is multidimensional. You may be
familiar with quadrature methods such as the midpoint or trapezoid rule from elementary calculus; the quadrature methods in MATLAB are close cousins. The
integrand is sampled at automatically selected points, and areas of approximating
shapes are used to estimate the area represented by the integral itself. The process
usually continues until the accuracy of the approximation is estimated to be less
than an optionally specified tolerance. The error estimate is not a mathematical
guarantee, but it can usually be treated with confidence.

MATLAB gives a few different quadrature variants (shown below), each of
which works best in certain situations. `quad` is best for general purposes unless
very high accuracy is desired, in which case `quadl` is likely to be more efficient.
The `quadgk` function is best for rapidly oscillating integrands, improper integrals,
and integration in the complex plane.

```
>> f = @(x) x.*sin(exp(x));
>> format long
>> [quad(f,1,6) quadl(f,1,6) quadgk(f,1,6)]'
ans =
  -0.340051345224177
  -0.340044091632578
  -0.340044091659739

>> [quad(f,1,8) quadl(f,1,8) quadgk(f,1,8)]'
Warning: Maximum function count exceeded.
> In quad at 106
Warning: Maximum function count exceeded.
> In quadl at 104

ans =
  -0.518113806710640
  -0.663567101898307
  -0.333649594278458

>> quadgk(@(x) 1./sqrt(x),0,1)
ans =
   1.999999999999763
```

Indefinite integration is normally a symbolic process of producing an
antiderivative, something that is not available through numerics. However, since
$\int_a^x f(t)\,dt$, where x is regarded as an independent variable, is an antiderivative of
$f(x)$, numerical quadrature can be used to create numerical values, though not a
formula, for any particular antiderivative. This is easily done with the `quad` family, but the result is quite inefficient, since a completely new quadrature is done
for each value of x. Instead, you should write the antiderivative as an initial-value
problem (IVP) and use one of the methods described in section 7.7.

MATLAB offers dblquad for cubature over two-dimensional rectangular domains and triplequad for integration over rectangular boxes in space. More general domains can be implemented straightforwardly using iterated integration, but in more than two dimensions this is probably much too slow in most cases.

7.7 Initial-value problems

MATLAB has an excellent library of solvers for IVPs in the form

$$\frac{dx}{dt} = f(t, x(t)), \quad a \le t \le b, \quad x(a) = x_0, \tag{7.1}$$

where x and f are understood to be vector valued. Any differential equation with higher-order derivatives present must first be written in this first-order form, as explained in most elementary books on differential equations and numerical analysis.

The primary division between IVP solvers is **stiff** versus **nonstiff**. Loosely speaking, stiff problems include phenomena that vary over widely different time scales, and they require methods that are more expensive for each time step taken, especially in the face of high-dimensional nonlinearity. The logical extreme of stiffness is the **differential-algebraic equation**, which has algebraic side constraints on the variables in addition to the dynamics.

A basic IVP solution takes the form

$$output = \text{odexxx}(\text{f}, \text{time}, \text{init})$$

The name of the function called might be ode45, ode113, ode15s, or one of the other solvers listed in the documentation. In terms of equation (7.1), the first input argument f is a callable function representing the mathematical function $f(t, x)$, the second argument time defines the interval $[a, b]$ for the independent variable t, and the input init defines the initial value x_0. Both the time input argument and the output can take different forms.

For example, consider the well-known predator-prey model

$$\frac{dx_1}{dt} = x_1 - \alpha x_1 x_2,$$
$$\frac{dx_2}{dt} = \beta x_1 x_2 - x_2.$$

This problem can be solved by first writing an M-file to represent the functions on the right-hand side. Note that the IVP solvers want the result $f(t, x)$ as a column vector:

```
function dxdt = predprey(t,x)

dxdt(1,1) = x(1) - 0.2*x(1)*x(2);
dxdt(2,1) = 0.5*x(1)*x(2) - x(2);

end
```

Here are several variations on solving the problem:

```
>> % auto-select times for output
>> [t,x] = ode45(@predprey, [0 20], [8;1]);
>> size(x)
ans =
   101      1

>> [ t(1:5)  x(1:5,:) ]
ans =
        0     1.0000
   0.0100     1.0001
   0.0200     1.0002
   0.0300     1.0005
   0.0400     1.0008

>> % manually select times for output
>> [t,x] = ode45(@predprey, [0:0.05:20], [8;1]);
>> size(x)
ans =
   401      2

>> [ t(1:5)  x(1:5,:) ]
ans =
        0     8.0000     1.0000
   0.0500     8.3198     1.1665
   0.1000     8.6363     1.3717
   0.1500     8.9445     1.6255
   0.2000     9.2375     1.9408

>> % delay selecting output times
>> sol = ode45(@predprey, [0 20], [8;1]);
>> deval(sol,t(1:5))
ans =
   8.0000     8.3198     8.6363     8.9444     9.2374
   1.0000     1.1665     1.3717     1.6255     1.9408
```

The internal solution process is not affected by whether or how you request output times. If you want to make solutions more (or less) accurate than the selected defaults, see the help on `odeset` or the online examples.

The IVP solvers offer lots of customization and additional features. Investigate online help if you are doing any more than the basics.

7.8   Boundary-value problems

A boundary-value problem (BVP) also starts with a first-order ODE, $y'(x) = f(x,y)$, for $a \le x \le b$, but unlike an IVP, it has side conditions involving two values of the solution: $g(y(a), y(b)) = 0$ for a vector-valued g. Existence and uniqueness theory is less general for BVPs than for IVPs, and the numerical solution process is also less straightforward. Here we just give a simple case study, the Allen–Cahn equation:

$$\epsilon u'' + u(1 - u^2) = 0, \quad 0 < x < 1; \qquad u(0) = -1, \quad u(1) = 1.$$

Both $u \equiv -1$ and $u \equiv 1$ are constant solutions of the differential equation, and this equation models a "phase change" between them. If ϵ is small, this change is abrupt. As with IVPs, the second-order differential equation must be cast in first-order vector form by introducing $y_1 = u$, $y_2 = u'$.

Three elements are needed to attempt a numerical solution: a function to define the ODE, a function to define the boundary conditions, and an initial guess to the solution. One way to organize these needs is to write a single M-file that returns them as output. For the Allen–Cahn example, with $\epsilon = 0.01$, we might use the following:

```
function [f,g,init] = mybvp(epsilon)
f = @(x,y) [ y(2); (y(1)^3-y(1))/epsilon ];
g = @(ya,yb) [ ya(1)+1; yb(1)-1 ];
x = linspace(0,1,10)';
yguess = @(x) [ -1+2*x; 2 ];
init = bvpinit(x,yguess);
```

The first output is the vector function $f(x, y)$. (In many problems this might be a handle to a subfunction.) The second is the vector function g that should equal zero when vectors $y(a)$ and $y(b)$ are substituted for ya and yb. In this case, only the first component of y appears at both boundaries. The third output is created by bvpinit, which in turn requires a function defining a guess for $y(x)$, which ought to satisfy the boundary conditions. Both the nodes and the values are potentially important to the solution process; in difficult problems, the guess may need to be quite similar to the true solution.

With these elements in place, we can call bvp4c or bvp5c to create the solution and use deval to evaluate it for a plot:

```
>> [f,g,init] = mybvp(1e-3);
>> soln = bvp4c(f,g,init)
soln =
            x: [1x143 double]
            y: [2x143 double]
           yp: [2x143 double]
       solver: 'bvp4c'

>> xp=0:0.01:1;  plot( xp, deval(soln,xp,1) )
```

The result is shown in Figure 7.3. The structure soln tells us that the original 10 nodes in x have been replaced by 143 unequally spaced ones, in order to represent the solution accurately.

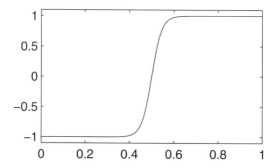

Figure 7.3. *Solution of the Allen–Cahn equation by* bvp4c.

If we attempt the same process with $\epsilon = 10^{-4}$, the solver appears to hang. The reason is that as ϵ shrinks, so does the width of the transition region around $x = 1/2$, and the linear initial guess for the solution is not sufficiently close to the correct shape. In this case one gets better results by using the solution with $\epsilon = 10^{-3}$ as the initial guess, in a crude form of a process called **continuation**. The syntax of bvp4c makes this easy—note the subtle change in the second line here:

```
>> [f,g,init] = mybvp(1e-4);
>> soln = bvp4c(f,g,soln);
```

7.9 Time-dependent partial differential equations

The numerical solution of partial differential equations (PDEs) is an enormous and difficult subject. In this section we show how some simple tools can be used to solve some problems in one space dimension plus time. While MATLAB does have an automatic facility for such problems called pdepe, here we present a semi-automated process known as semidiscretization or **the method of lines**. In the method of lines one discretizes space and time independently. At a practical level this means the original PDE is replaced by a (typically large) system of ODEs. In principle, and often in practice too, one can solve the ODE system using the IVP solvers from section 7.7.

For example, consider the problem

$$u_t = 0.2u_{xx} + u, \quad -1 \le x \le 1, \qquad u(-1,t) = u(1,t) = 0. \qquad (7.2)$$

This problem mixes exponential growth (via $u_t = u$) with diffusion. To solve this problem, we divide the interval $[-1,1]$ into n intervals of width $h = 2/n$, having endpoints x_i for $i = 0,\ldots,n$. We represent the function $u(x,t)$ at a fixed

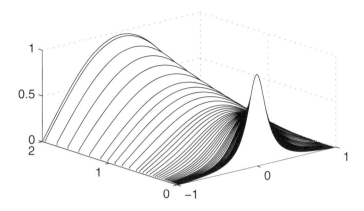

Figure 7.4. *Method of lines result for a PDE in one dimension.*

time t as a vector of its values $u_i(t)$, $i = 1,\ldots,n-1$, with the understanding that $u_0(t) = u_n(t) = 0$ for all time due to the boundary conditions. The differentiation operation u_{xx} is replaced by the matrix-vector multiplication

$$\frac{1}{h^2}\begin{bmatrix} -2 & 1 & & & \\ 1 & -2 & 1 & & \\ & \ddots & \ddots & \ddots & \\ & & 1 & -2 & 1 \\ & & & 1 & -2 \end{bmatrix}\begin{bmatrix} u_1 \\ u_2 \\ \vdots \\ u_{n-2} \\ u_{n-1} \end{bmatrix}. \tag{7.3}$$

The result is a linear IVP for the semidiscrete values $u_i(t)$, which can be solved using the built-in methods. The problem can be attacked with the following remarkably short script, whose result is shown in Figure 7.4:

```
n = 100; h = 2/n;                  % n intervals, width 2/n
x = -1 + h*(1:n-1)';               % node locations
D2 = toeplitz([-2 1 zeros(1,n-3)]/h^2); % diff matrix
f = @(t,u) 0.2*D2*u + u;           % discretized du/dt
u0 = (1-x.^2)./(1+50*x.^2);        % initial condition
[t,u] = ode15s(f,[0 2],u0);        % solve
waterfall(x,t,u)
```

This script is far more effective than trying to code a "classical" method such as Crank–Nicolson. Observe that small time steps were taken initially because of the large derivatives in the solution, and that much larger steps became possible once the solution smoothed out.

Exercises

7.1. Write a function or script that solves the linear system $A^k x = b$ for a square matrix A and positive integer k, using an LU factorization of A and never explicitly computing the matrix A^k. (Hint: Interpret the problem as solving k linear systems sequentially.)

7.2. Suppose A is a full-rank 6×4 matrix. Which of the following is (in exact arithmetic) equal to an identity matrix: A\A or A/A? Why is the other case something other than the identity?

7.3. Examine the eigenvalues of the family of matrices

$$D_N = -N^2 \begin{bmatrix} -2 & 1 & 0 & 0 & \cdots & 1 \\ 1 & -2 & 1 & 0 & \cdots & 0 \\ 0 & 1 & -2 & 1 & \cdots & 0 \\ & & \ddots & \ddots & \ddots & \vdots \\ 0 & 0 & \cdots & 1 & -2 & 1 \\ 1 & 0 & 0 & \cdots & 1 & -2 \end{bmatrix},$$

where D_N is $N \times N$, for growing values of N, for example, $N = 32, 64, 128$. The smallest eigenvalues converge to integer multiples of a simple number.

7.4. Verify, for a random 10×5 matrix, the identity

$$\|A\|_F^2 = \sum_{i=1}^{K} \sigma_i^2,$$

where $K = \min\{m, n\}$, $\{\sigma_1, \ldots, \sigma_K\}$ are the singular values of A, and $\|\cdot\|_F$ is the Frobenius matrix norm.

7.5. A classical problem of applied mathematics is to find the zeros of Bessel's function $J_\nu(x)$ for fixed values of the index ν. Find all the zeros of $J_{1/2}(x)$ for $x \in [0, 10]$.

7.6. Lambert's W function is defined as the inverse of the function $f(x) = xe^x$. It has no simple analytic expression. Write a function W = lambert (x) that evaluates Lambert's W at any value of $x > 0$. (Hint: Solve the expression $x = ye^y$ for y given x, using fzero.)

7.7. Find the value of $x \in [0, 1]$ that minimizes the largest eigenvalue of the matrix $A(x) = xM + (1-x)P$, where M is a 5×5 magic square and P is a 5×5 Pascal matrix.

7.8. Consider the census population data fits in section 7.5. Show that by setting $c_1 = 0$ in the proposed model, a linear relationship results between t and the log of the population. Hence use `polyfit` to find c_2 and c_3 in the model, and check the quality of fit graphically. (See also section 5.1.2.)

7.9. Write a function

$$\texttt{function F = indefint(f,a,b,Fa)}$$

that returns a callable F such that $F(x)$ evaluates for $x \in [a,b]$ the antiderivative of f defined uniquely by $F(a) = \texttt{Fa}$. (A call to `ode45`, followed by an anonymous function wrapper around a call to `deval`, is a straightforward solution.)

7.10. The quadrature routines require an integrand that can be evaluated for any value of the independent variable. A function given only by some of its values at discrete points must first be interpolated before quadrature can be applied. Write a function

$$\texttt{function Q = interpquad(x,y)}$$

that applies `quad` over the interval $[\texttt{x(1)},\texttt{x(end)}]$ to the piecewise cubic interpolant computed by `interp1` for the data points defined by vectors x and y. (Note: If instead you use a piecewise linear interpolant, the result is the Trapezoid Rule for numerical quadrature.)

7.11. Consider the IVP $x'(t) = tx^2$, $x(0) = 1$. Solve this problem using `ode45` for $0 \le t \le 1$, and plot the result. Then try solving again for $0 \le t \le 2$. How do you interpret what happens? This IVP is a separable equation with an easy solution, so you can check your guess.

7.12. Solve the BVP

$$y''' = -2e^{-3y} + 4(1+x)^{-3}, \quad 0 \le x \le 1,$$

subject to $y(0) = 0$, $y'(0) = 1$, $y(1) = \ln 2$. Compare to the exact solution, $y(x) = \ln(1+x)$.

7.13. Solve the linear advection-diffusion equation,

$$u_t = 0.2u_{xx} - u_x, \quad -1 \le x \le 1, \quad 0 \le t \le 2,$$

with the boundary conditions $u(-1,t) = u(1,t) = 0$ and the initial condition from the code on page 89. In place of u_x, use the centered finite difference matrix

$$\frac{1}{2h}\begin{bmatrix} 0 & 1 & 0 & & & \\ -1 & 0 & 1 & & & \\ & \ddots & \ddots & \ddots & & \\ & & -1 & 0 & 1 \\ & & & -1 & 0 \end{bmatrix}\begin{bmatrix} u_1 \\ u_2 \\ \vdots \\ u_{n-2} \\ u_{n-1} \end{bmatrix}.$$

7.14.   (a) Solve the PDE $u_t = 0.2u_{xx} + u^2$, $-1 \le x \le 1$, $u(-1,t) = u(1,t) = 0$, with the same initial condition as the code on page 89, for $0 \le t \le 2$.

       (b) Disable the plotting command in the code, multiply the initial condition by 10, and try to solve again. You should find that the solution reaches a singularity in finite time (as does $u_t = u^2$).

Afterword

What now?

If you have gotten this far, typing in examples and giving an honest effort to many of the exercises, then you are well beyond a beginning MATLAB programmer. From here, any further learning about MATLAB should be driven by your own needs. Are you creating a module or toolbox for other MATLAB users? Maybe you should consider an object-oriented programming model. Want to create a friendly interface for nontechnical users who may not even have access to MATLAB? Try a graphical user interface and the MATLAB Compiler. Need to incorporate Fortran legacy code? See the documentation on "external interfaces." Or you might delve into one of the many toolboxes for specific application areas.

The point is that you now should understand the structures on which MATLAB programming is based. Greater fluency is mainly a matter of increasing your vocabulary and gaining experience. I encourage you, before writing a block of code to accomplish a generic task, to take a few moments to look around the documentation and matlabcentral.com. As Samuel Johnson said, "A man will turn over half a library to write one book."

Don't forget too about the mathematics underlying so many of the MATLAB functions you might use. This body of mathematics, most often collected under the rubric of numerical analysis, is at least as responsible as hardware improvements for the breathtaking increases of scope in the problems scientists and engineers have been able to solve over the last several decades. Numerical analysis is not just useful but often strikingly beautiful. By learning MATLAB—not just using it—you can learn about some of the most exciting developments and triumphs of this important field.

Newton stood on the shoulders of giants, but you have the opportunity to order them around and put them to work. Happy coding!

Toby Driscoll
January 2009

Index